国家自然科学基金项目(61602217)、江苏高校"青蓝工程"优秀青年骨干教师项目(苏教师函〔2021〕11号)等基金项目资助出版

基于拉曼光谱的乳制品质量智能判别技术研究

张正勇 著

东南大学出版社
SOUTHEAST UNIVERSITY PRESS

·南京·

图书在版编目（CIP）数据

基于拉曼光谱的乳制品质量智能判别技术研究 / 张正勇著. -- 南京：东南大学出版社，2024.12.
ISBN 978-7-5766-1723-8

Ⅰ. TS252.7

中国国家版本馆 CIP 数据核字第 2024LG5040 号

基于拉曼光谱的乳制品质量智能判别技术研究

Jiyu Laman Guangpu De Ruzhipin Zhiliang Zhineng Panbie Jishu Yanjiu

著　　者	张正勇
出版发行	东南大学出版社
社　　址	南京市四牌楼 2 号（邮编：210096）
出 版 人	白云飞
网　　址	http://www.seupress.com
策划编辑	孙松茜
责任编辑	孙松茜
责任校对	韩小亮
封面设计	王　玥
责任印制	周荣虎
经　　销	全国各地新华书店
印　　刷	广东虎彩云印刷有限公司
开　　本	700 mm×1000 mm　1/16
印　　张	13.5
字　　数	272 千字
版　　次	2024 年 12 月第 1 版
印　　次	2024 年 12 月第 1 次印刷
书　　号	ISBN 978 - 7 - 5766 - 1723 - 8
定　　价	88.00 元

（本社图书若有印装质量问题，请直接与营销部联系。电话：025 - 83791830）

前 言
PREFACE

乳及乳制品作为大众生活必需品，在人们的日常生产生活过程中处于特殊地位。随着乳制品营养健康的理念愈发深入人心，乳制品人均消费量已被列为衡量一个国家的人民生活水平的重点指标之一，乳制品消费呈现出持续刚性增长态势及多样化需求趋势。在乳业发展过程中也历经了波折，不过总体上奶牛存栏数、乳制品产量、乳制品消费能力及企业的生产能力近年来都得到了极大的提高。据国家统计局发布的数据显示，2023年我国牛乳产量4 197万吨，增长6.7%。据《中国食品报》的报道显示，2023年我国乳制品行业市场规模已经突破5 000亿元。乳制品的安全与质量控制的重要性毋庸置疑，然而近年来，乳制品质量安全相关问题仍时有发生，诸如奶制品污染事件，假奶粉事件以及海外输入劣质奶粉事件等，这些问题引起了公众和国家层面的高度重视。在《国务院办公厅关于推进奶业振兴保障乳品质量安全的意见》(国办发〔2018〕43号)、《"十四五"奶业竞争力提升行动方案》(农牧发〔2022〕8号)等文件中，均明确提出要强化乳品质量安全监管。为实现这一目标，需要不断创新监管方式，发展基于大数据的智能统计分析技术应用，深入开展乳制品交叉学科研究，不断完善乳制品质量控制手段，持续强化科技支撑技术体系，助力推动乳业高质量发展。

在传统的乳制品质量分析技术研发领域，常关注的是专家感官鉴定法、成分分析法，以实现整体和特定成分的判别分析，但也存在品鉴专家主观不确定性，特定成分判别的片面性，尤其是当面对相似样品以次充好时难以做到判定具有针对性、有效性，故而迫切需要发展面对这一挑战的判别新技术。近年来，基于拉曼光谱的食品质量安全智能学习算法研究日益引起科研人员的广泛关注，原因在于拉曼光谱可快速、高效地表征食品样品的分子振动信息，并具有样品用量少、可无损检测、仪器易于便携化、含水样品可直接测试等多种优点，智能学习算法具有运算速度快、判别准确率高、结果客观等特点，两者的结合为乳制品质量问题的判别分析提供了新的研究方案，为此很有必要围绕乳制品质量智能判别开展和总结相关研究工作。

本书的结构分为6章，第1章围绕乳制品相关质量分析研究进展，第2章围绕乳制品拉曼光谱数据预处理方法，第3章围绕乳制品拉曼光谱数据特征提取，第4章围绕乳制品高维拉曼光谱数据构建，第5章围绕乳制品拉曼光谱相似性度量，第6章围绕乳制品拉曼光谱智能判别技术，依次进行了梳理分析，通过相关实验研究较为全面地探讨了各个技术细节。

本书的相关研究工作得到了国家自然科学基金(61602217)，江苏高校"青蓝工程"优秀青年骨干教师项目(苏教师函〔2021〕11号)等基金资助，在此表示衷心的感谢。

由于作者水平有限，尽管撰写过程中尽了最大的努力，但在内容上仍可能存在局限性和不足之处，恳请参阅本书的读者提出宝贵意见，以利于作者进一步开展相关研究工作，不断改进和完善相关研究技术。

作　者

2024年6月

目录
CONTENTS

第1章 乳制品质量分析研究进展 ……………………………………………… 1
  1.1 乳制品相关研究进展 …………………………………………………… 1
  1.2 乳制品相关质量评价技术研究进展 …………………………………… 5
  1.3 基于拉曼光谱的乳制品相关研究进展 ………………………………… 9

第2章 乳制品拉曼光谱数据的预处理研究 …………………………………… 12
  2.1 乳粉拉曼光谱表征数据的标准化与降噪处理研究 …………………… 12
  2.2 基于小波变换的乳制品智能鉴别技术优化研究 ……………………… 21

第3章 乳制品拉曼光谱数据特征提取研究 …………………………………… 28
  3.1 基于拉曼光谱化学特征提取的乳制品质量判别研究 ………………… 28
  3.2 基于拉曼光谱特征提取与融合的乳制品统计判别 …………………… 40
  3.3 基于机器学习算法和拉曼光谱的乳制品鉴别特征分析 ……………… 56

第4章 乳制品高维拉曼光谱数据的构建研究 ………………………………… 71
  4.1 激光微扰二维相关拉曼光谱用于乳制品质量判别 …………………… 71
  4.2 基于高维拉曼光谱的鲜奶制品表征研究 ……………………………… 81

第5章 基于相似性度量的乳制品质量判别技术研究 ………………………… 97
  5.1 基于拉曼光谱相关系数的乳粉鉴别技术研究 ………………………… 97
  5.2 基于拉曼光谱相似性度量的乳粉高维表征及判别分析 ……………… 103

第 6 章　基于智能判别算法的乳制品质量智能判别技术研究 …………… 113
　6.1　基于拉曼光谱与 k 近邻算法的酸奶鉴别研究 …………………………… 113
　6.2　基于支持向量机算法的乳制品分类识别技术研究 ……………………… 121
　6.3　基于拉曼光谱与随机森林算法相结合的豆乳粉品牌识别 ……………… 128
　6.4　基于拉曼光谱和概率神经网络的乳酪制品快速鉴别 …………………… 139
　6.5　拉曼光谱结合模式识别算法的乳制品智能判别与参数优化 …………… 154
　6.6　基于拉曼光谱和极限学习机的乳酪制品表征参数优化 ………………… 162

参考文献 ……………………………………………………………………………… 181

第 1 章　乳制品质量分析研究进展

乳及乳制品是关系国民健康的重要营养来源，乳制品产业又是国民经济的重要组成部分，其质量安全监管始终是国家和民众关注的重点。近年来，围绕乳制品千亿产业、产品开展的相关研究工作异常丰富，据 Web of Science 数据库中检索到的文献发表量可以发现每年相关研究报道均超过 1 万篇，显示出乳制品产业、产品这一研究领域受到了广泛的关注，相关科研成果丰硕，相关技术进步迭代趋势明显，接下来将依次从乳制品相关研究发展形势、质量评价技术动态变化趋势、拉曼光谱相关技术演进态势展开进一步分述。

1.1　乳制品相关研究进展

乳制品相关研究涉及范畴相当宽泛，从乳源分类来看，包括了牛乳、人乳以及山羊乳、骆驼乳等特色乳，目前牛乳依然是市场主体。从产品供给形式来看，包括了鲜乳、巴氏杀菌乳、发酵乳、乳粉、乳酪、乳饮料等。从质量安全风险来看，包括了非法添加物、微生物污染、农兽药重金属添加剂残留、以次充好等。从质量安全检测技术来看，包括了色谱、质谱、光谱、生物免疫技术等[1]。以下将从乳制品营养健康与生产工艺、生物合成制备与调控、质量安全检测技术、生物活性菌技术、功能乳品、产业绿色可持续发展等方面对最新研究进展进行一定的梳理阐述。

（1）乳制品营养健康与生产工艺方面，乳制品是理想的营养来源，被认为是含有碳水化合物、蛋白质、脂肪、矿物质和维生素的平衡组合，此外，还在提供必要的有益菌群方面发挥着关键作用，可能有助于加强新生哺乳动物的免疫系统，Bodor 等研究了罗马尼亚不同乳制品品种的营养数据，经统计检验发现在各类型乳制品产品中脂肪和能量值之间存在显著的正相关，黄油中的脂肪和蛋白质含量之间存在显著的负相关，乳酪中蛋白质和脂肪酸含量密切相关[2]。Kaplan 等研究

---

[1] 钮伟民. 乳及乳制品检测新技术[M]. 北京：化学工业出版社，2012：1-280.
[2] Bodor K，Tamási B，Keresztesi Á，et al. A comparative analysis of the nutritional composition of several dairy products in the Romanian market[J]. Heliyon，2024，10(11)：e31513.

指出 A2 牛奶含有 A2β-酪蛋白而不是 A1β-酪蛋白,是同等营养含量的普通牛奶的最佳替代品,且不会引起任何胃肠道不适或普通牛奶消费中出现的更严重的问题[①]。Salvo 等研究评估了驴奶和骆驼奶中的不同化合物,分析了它们的生物分子特征和对人类健康的潜在益处,研究指出驴奶和骆驼奶的生物活性产品可能是控制几种疾病的良好候选者,也是婴儿牛奶蛋白过敏的良好替代品[②]。Ho 等研究了骆驼奶的成分、生物活性化合物和热稳定性,并指出牛奶和骆驼奶成分的差异使得牛奶产品的生产工艺不适合骆驼奶产品,文章进一步讨论了各种骆驼奶产品生产的技术困难与改进方向[③]。da Cunha 等系统总结了膜法这一新兴的低能耗技术用于牛奶的浓缩、纯化,并指出超滤、反渗透工艺目前的挑战主要在于控制受系统操作条件如污垢和极化等现象的影响[④]。

(2)乳制品生物合成制备与调控方面,Yang 等研究指出母乳低聚糖这一公认的婴儿健康关键功能生物分子,其最主要的核心结构——乳糖-N-四糖具有良好的生理作用,包括益生元特性、抗粘连、抗菌活性和抗病毒作用。乳糖-N-四糖的安全性已经过评估,并已经作为一种功能成分在商业上添加到婴儿配方奶粉中,高效的生物生产尤其是基于代谢工程的微生物合成在乳糖-N-四糖的大规模生产中显示出明显的优势[⑤]。Meng 等指出母乳低聚糖是由乳糖通过逐步糖基化延伸或修饰而酶促产生的,每种糖基化都需要一种特定的糖基转移酶和相应的核苷酸糖供体,文章介绍了一种一锅多酶法,并描述了母乳低聚糖合成的酶级联路线的设计原理[⑥]。Kellman 等采用整合聚糖和核糖核酸(Ribonucleic Acid,RNA)表达数据的系统生物学框架来构建母乳低聚糖生物合成网络,并预测所涉及的糖基转移酶,提出了母乳低聚糖的延伸、分支、岩藻糖基化和唾液酸化的候选

---

① Kaplan M, Baydemir B, Günar B B, et al. Benefits of A2 milk for sports nutrition, health and performance[J]. Frontiers in Nutrition, 2022, 9: 935344.

② Salvo E D, Conte F, Casciaro M, et al. Bioactive natural products in donkey and camel milk: A perspective review[J]. Natural Product Research, 2023, 37(12): 2098-2112.

③ Ho T M, Zou Z Z, Bansal N. Camel milk: A review of its nutritional value, heat stability, and potential food products[J]. Food Research International, 2022, 153: 110870.

④ da Cunha T M P, Canella M H M, da Silva Haas I C, et al. A theoretical approach to dairy products from membrane processes[J]. Food Science and Technology, 2022, 42: e12522.

⑤ Yang L H, Zhu Y Y, Zhang W L, et al. Recent progress in health effects and biosynthesis of lacto-N-tetraose, the most dominant core structure of human milk oligosaccharide[J]. Critical Reviews in Food Science and Nutrition, 2024, 64(19): 6802-6811.

⑥ Meng J W, Zhu Y Y, Wang H, et al. Biosynthesis of human milk oligosaccharides: Enzyme cascade and metabolic engineering approaches[J]. Journal of Agricultural and Food Chemistry, 2023, 71(5): 2234-2243.

基因,为母乳低聚糖生物合成提供了必要的分子基础①。

(3)质量安全检测技术方面,Nagraik等针对牛奶掺杂物检测进行了系统总结,涉及牛奶中故意混合的水、植物和动物脂肪、外来蛋白质和化学成分(即三聚氰胺、尿素、福尔马林、洗涤剂、硫酸铵、硼酸、烧碱、苯甲酸、水杨酸、过氧化氢和糖)等,为此,研究人员开发了各种检测牛奶掺杂物的方法,包括采用高效液相色谱法、气相色谱法和质谱法对不同的牛奶掺杂物进行了选择性鉴定和检测。免疫技术,如酶联免疫吸附试验(Enzyme Linked Immunosorbent Assay,ELISA)和各种基于脱氧核糖核酸(Deoxyribo Nucleic Acid,DNA)的方法,如聚合酶链式反应(Polymerase Chain Reaction,PCR),用于一些常见牛奶掺杂物的特异性检测;光谱方法,即傅里叶变换红外光谱(Fourier Transform Infrared Spectroscopy,FTIR)和近红外光谱(Near Infrared,NIR)与化学计量学相结合,提高了掺杂物检测系统能力;电子鼻和电子舌等设备也是牛奶和其他食品掺杂物检测中使用的一种技术手段,生物传感器亦可用于牛奶掺杂物的快速实时检测等②。Liu等建立了一种基于流式细胞术的快速检测牛奶和奶粉中金黄色葡萄球菌的方法,该方法可以在6小时内检测到少量金黄色葡萄球菌细胞,牛奶中的检测限为7.50个细胞/mL,奶粉中的检测限为8.30个细胞/g③。Zhu等系统总结了基于色谱、光谱、介电性质和传感器的牛奶中主要成分的快速检测技术,并分析此类快检技术的潜在优势和局限性④。

(4)生物活性菌技术方面,Yadav等研究了不同乳成分对肠道微生物组与婴儿健康的调节关系,研究发现乳脂肪球膜蛋白可以直接与益生菌相互作用,通过胃肠道转运影响其存活和黏附,而膜磷脂增加了益生菌在肠道中的停留时间。牛奶中的共生细菌是建立婴儿肠道定植的初始接种物,而低聚糖则促进有益微生物的增殖⑤。相较传统的发酵牛奶产品,Shori等则关注了大豆、椰子等非乳制品,

---

① Kellman B P, Richelle A, Yang J Y, et al. Elucidating Human Milk Oligosaccharide biosynthetic genes through network-based multi-omics integration[J]. Nature Communications, 2022, 13: 2455.
② Nagraik R, Sharma A, Kumar D, et al. Milk adulterant detection: Conventional and biosensor based approaches: A review[J]. Sensing and Bio-Sensing Research, 2021, 33: 100433.
③ Liu S Y, Wang B, Sui Z W, et al. Faster detection of Staphylococcus aureus in milk and milk powder by flow cytometry[J]. Foodborne Pathogens and Disease, 2021, 18(5): 346 – 353.
④ Zhu Z Z, Guo W C. Recent developments on rapid detection of main constituents in milk: A review[J]. Critical Reviews in Food Science and Nutrition, 2021, 61(2): 312 – 324.
⑤ Yadav M, Kapoor A, Verma A, et al. Functional significance of different milk constituents in modulatingthe gut microbiome and infant health[J]. Journal of Agricultural and Food Chemistry, 2022, 70(13): 3929 – 3947.

研究探讨了益生菌在植物奶中的生长和活力影响因素，涉及奶成分、发酵过程、益生菌类型、储存时间和温度、酸度和包装等①。包秋华等应用拉曼光谱技术对比分析了德氏乳杆菌保加利亚亚种ND02正常及其非可培养状态细胞的表观形态和细胞内部核酸、脂类以及蛋白类大分子化合物的变化情况，在单细胞水平进行了两个菌种细胞内的生物大分子检测分析②。

（5）功能乳品技术方面，Wang等研究显示牛奶来源的生物活性肽具备调节心血管、免疫、消化和神经系统作用，乳源生物活性肽在食品和生物医学领域受到了广泛关注，尤其是在特殊医学用途食品领域，成为一种新的患者营养管理解决方案③。Ma等研发了一种基于高效液相色谱-紫外光谱准确鉴定特殊医学用途婴儿配方奶粉中谷蛋白的方法，谷蛋白是一种潜在的过敏原，可导致婴儿过敏、乳糜泻和糖尿病，甚至可能导致4个月以下婴儿的自身免疫性疾病，该方法的谷蛋白溶液检测限为 $9.7~\mu g/mL$，$R^2 = 0.9909$④。Huang等研究建立了一种基于同位素稀释液相色谱-串联质谱的测定乳及乳制品中维生素B-12的方法，该方法的检测限和定量限分别为 $0.5~\mu g/kg$ 和 $1.0~\mu g/kg$，实现了婴儿配方奶粉、特殊医学用途处方食品等不同类型牛奶和乳制品中维生素B-12的测定⑤。

（6）乳制品产业绿色可持续发展方面，Canavari等通过对意大利消费者进行调查研究，发现他们普遍愿意为标有低碳足迹标签的牛奶付费，绿色营销和相关的可持续标签是向消费者传达更可持续商业模式信息的重要手段。此外，消费者的支付意愿还取决于消费者对气候变化的重视程度、价格敏感性以及收入⑥。Gao等的研究发现大多数中国消费者并不清楚地理解可持续性的含义，并且缺乏

---

① Shori A B, Al Zahrani A J. Non-dairy plant-based milk products as alternatives to conventional dairy products for delivering probiotics[J]. Food Science and Technology，2022，42：e101321.

② 包秋华，马学波，任艳，等. 应用拉曼光谱对比分析德式乳杆菌保加利亚亚种ND02及其VBNC态细胞成分[J]. 食品科学，2022，43(10)：114-118.

③ Wang L L, Shao X Q, Cheng M, et al. Mechanisms and applications of milk-derived bioactive peptides in Food for Special Medical Purposes[J]. International Journal of Food Science & Technology，2022，57(5)：2830-2839.

④ Ma M J, Wang C F, Sun H Y, et al. Precise and efficient HPLC-UV identification of rice glutelin in infant formulas for special medical purposes[J]. International Journal of Food Science & Technology，2023，58(7)：3769-3780.

⑤ Huang B F, Zhang J S, Wang M L, et al. Determination of vitamin B12 in milk and dairy products by isotope-dilution liquid chromatography tandem mass spectrometry[J]. Journal of Food Quality，2022，2022：7649228.

⑥ Canavari M, Coderoni S. Green marketing strategies in the dairy sector: Consumer-stated preferences for carbon footprint labels[J]. Strategic Change-Briefings in Entrepreneurial Finance，2019，28(4)：233-240.

关于可持续食品生产的知识。消费者愿意为可持续牛奶支付的溢价约为40%。没有意识到可持续生产与食品质量之间联系的消费者购买可持续牛奶的意愿明显较低[①]。Wanniatie等进行了传统和有机奶牛场的比较,以及评估了有机牛奶在营养和污染物方面的质量。研究发现有机农业是对生产工艺和产品质量有特殊要求的高端市场生产体系,对管理资质要求很高。有机牛奶不含农药残留、抗生素和其他污染物,较之传统牛奶,有机牛奶的共轭亚油酸含量较高[②]。Samarra等研究提出氧化脂质作为一组衍生自各种多不饱和脂肪酸的氧化代谢产物,可能是有机牛奶评估的有前途的生物标志物,用以区分有机牛奶和传统牛奶[③]。

1.2 乳制品相关质量评价技术研究进展

乳制品是富含多种化学组分的复杂混合体系,在质量评价分析领域所建立的方法有很多,此部分将评价方法主要分为专家感官鉴别法和仪器分析法展开进一步的论述。专家感官鉴别法即利用专业人员敏锐的感受器官进行样品的色、香、味和外观形态度量,实现待测样品质量评价,如《食品安全 国家标准 生乳》(GB 19301—2010)明确规定了生乳的色泽应为乳白色或微黄色,滋味和气味要有乳固有的香味,无异味,组织状态要呈现出均匀一致液体,无凝块、无沉淀、无正常视力可见异物[④]。谢琳等探讨了使用感官鉴别法中的三点检验法和分类检验法,针对不同热处理工艺生产的纯牛乳制品加以区分的可行性[⑤]。不过这类方法易受环境、身体状态、情感因素等影响,主观性较强。仪器分析法可细分为仿生仪器分析法和成分分析法。其中,仿生仪器分析法主要是利用仿生设备如电子鼻、电子舌、电子眼模拟感官传感收集样品响应信号。Zeng等采集了四种不同风味的酸奶样品的电子鼻数据,与线性判别分析、逻辑回归、支持向量机和决策树算法相结合研

---

① Gao Z F, Li C G, Bai J F, et al. Chinese consumer quality perception and preference of sustainable milk[J]. China Economic Review, 2020, 59: 100939.
② Wanniatie V, Sudarwanto M B, Purnawarman T, et al. Milk quality from organic farm[J]. Wartazoa-Buletin Ilmu Peternakan dan Kesehatan Hewan Indonesia, 2017, 27(3): 125-134.
③ Samarra I, Masdevall C, Foguet-Romero E, et al. Analysis of oxylipins to differentiate between organic and conventional UHT milks[J]. Food Chemistry, 2021, 343: 128477.
④ 中华人民共和国卫生部.食品安全国家标准 生乳: GB 19301—2010[S].北京:中国标准出版社, 2010.
⑤ 谢琳,王晓君,刘彭,等.感官检验法鉴别市售纯牛乳制品热处理方式的探讨[J].乳业科学与技术, 2008,31(4):179-182,188.

究建立了香气类型分类模型，实现了酸奶香气特征的快速质量判别[1]。Perez-Gonzalez等开发了一种基于银纳米粒子和酶的灵敏度和选择性增强的电子舌传感器阵列，可对感兴趣的牛奶成分（如葡萄糖、半乳糖、乳糖和尿素）的反应有着高的敏感性，进一步结合主成分分析及支持向量机算法可实现检测信号与物理化学参数之间相关性模型的建立[2]。Grassi等系统总结了电子眼、电子鼻、电子舌和数据分析技术相结合在乳制品货架期评价中的应用情况，揭示出电子传感系统在新鲜食品货架寿命评估方面的巨大潜力，同时也指出该类仿生设备面临着环境因素、老化、污染物吸附所致的传感器漂移等挑战[3]。成分分析法主要借助各类检验检测仪器，进行乳制品样品物理特性值、化学特征成分、生物基因序列等各类质量评价指标的定性、定量分析。Wang等研究了山羊乳制品和牛乳制品的低聚糖图谱差异，检测到27种低聚糖，并结合主成分分析法用于样品分类，鉴定出的乳糖-N-三糖可作为一种潜在的生物标志物，用于山羊乳与牛乳、山羊初乳粉与牛初乳粉以及山羊乳糖粉与牛乳糖粉的区分判别[4]。Sun等使用单分子实时测序技术，对西藏林芝市不同县牦牛奶、藏黄牛乳及其发酵产物中乳酸菌的群落组成进行了研究，发现德氏乳杆菌（36.17％）、嗜热链球菌（19.46％）和乳酸乳球菌（18.33％）为优势菌种，并考察了不同地点、不同牛奶类型和不同海拔高度因素对菌落组成的影响[5]。Shishov等研发了一种深共晶溶剂，可有效地液相微萃取乳制品中的三聚氰胺，然后通过高效液相色谱法进行测定。在最佳条件下，三聚氰胺测定的浓度范围为0.1至30 mg/L，检测限为0.03 mg/L[6]。此类研究实现了乳制品中多种特定物质的定性、定量分析，是目前乳制品质量分析的主要技术手段。但是，

---

[1] Zeng H, Han H Y, Huang Y D, et al. Rapid prediction of the aroma type of plain yogurts via electronic nose combined with machine learning approaches[J]. Food Bioscience, 2023, 56: 103269.

[2] Perez-Gonzalez C, Salvo-Comino C, Martin-Pedrosa F, et al. A new data analysis approach for an AgNPs-modified impedimetric bioelectronic tongue for dairy analysis[J]. Food Control, 2024, 156: 110136.

[3] Grassi S, Benedetti S, Casiraghi E, et al. E-sensing systems for shelf life evaluation: A review on applications to fresh food of animal origin[J]. Food Packaging and Shelf Life, 2023, 40: 101221.

[4] Wang H Y, Zhang X Y, Yao Y, et al. Oligosaccharide profiles as potential biomarkers for detecting adulteration of caprine dairy products with bovine dairy products[J]. Food Chemistry, 2024, 443: 138551.

[5] Sun Y, Zhao L X, Cai H Y, et al. Composition and factors influencing community structure of lactic acid bacterial in dairy products from Nyingchi Prefecture of Tibet[J]. Journal of Bioscience and Bioengineering, 2023, 135(1): 44-53.

[6] Shishov A, Nizov E, Bulatov A. Microextraction of melamine from dairy products by thymol-nonanoic acid deep eutectic solvent for high-performance liquid chromatography-ultraviolet determination[J]. Journal of Food Composition and Analysis, 2023, 116: 105083.

也面临着一些挑战,主要包括:(1)乳制品待测成分繁多,包括营养成分、有害微生物、生物毒素、抗生素、非法添加物、农药/兽药残留以及甜味剂、防腐剂等,仅为假冒蛋白质而报道使用的非法添加物[1][2],即有三聚氰胺[3]、双氰胺[4]、尿素[5]、硫酸铵[6]、皮革水解物[7]等层出不穷,造成单指标成分测试评价相对滞后,难以有效判别及控制未知风险。(2)针对组分近似的样品,利用个别成分予以判别存在不确定性。当非法人员使用低质量/低价格/易混的产品冒充高质量/高价格产品以谋取差价,其各项指标可能均符合国家质量安全标准,成为以质论价、公平合理的乳品市场购销秩序形成的一大制约,因此迫切需要研究发展新型质量判别技术[8]。(3)此类方法常常是建立在单个图谱数据中的峰值分析基础上,如基于朗伯比尔定律构建标准曲线,易造成实际使用中信息分析孤立、片面,图谱数据信息利用率相对偏低。

近年来,机器学习、数据挖掘等信息技术可有效提高检测表征数据综合利用效率、智能化判别水平,成为质量分析快速检测技术研究的重要趋势。Huang 等利用拉曼光谱结合偏最小二乘回归和反向传播人工神经网络,实现了针对牛奶酸

---

[1] Tian H X, Chen S, Li D, et al. Simultaneous detection for adulterations of maltodextrin, sodium carbonate, and whey in raw milk using Raman spectroscopy and chemometrics[J]. Journal of Dairy Science, 2022, 105(9): 7242-7252.

[2] Duan Y F, Chen Y H, Lei M K, et al. Hybrid silica material as a mixed-mode sorbent for solid-phase extraction of hydrophobic and hydrophilic illegal additives from food samples[J]. Journal of Chromatography A, 2022, 1672: 463049.

[3] Sereshti H, Mohammadi Z, Soltani S, et al. Synthesis of a magnetic micro-eutectogel based on a deep eutectic solvent gel immobilized in calcium alginate: Application for green analysis of melamine in milk and dairy products[J]. Talanta, 2023, 265: 124801.

[4] Nanayakkara D, Prashantha M A B, Fernando T L D, et al. Detection and quantification of dicyandiamide (DCD) adulteration in milk using infrared spectroscopy: A rapid and cost-effective screening approach[J]. Food and Humanity, 2023, 1: 1472-1481.

[5] Kumar V, Kaur I, Arora S, et al. Graphene nanoplatelet/graphitized nanodiamond-based nanocomposite for mediator-free electrochemical sensing of urea[J]. Food Chemistry, 2020, 303: 125375.

[6] Nieuwoudt M K, Holroyd S E, McGoverin C M, et al. Raman spectroscopy as an effective screening method for detecting adulteration of milk with small nitrogen-rich molecules and sucrose[J]. Journal of Dairy Science, 2016, 99(4): 2520-2536.

[7] Dong Y L, Yan N, Li X, et al. Rapid and sensitive determination of hydroxyproline in dairy products using micellar electrokinetic chromatography with laser-induced fluorescence detection[J]. Journal of Chromatography A, 2012, 1233: 156-160.

[8] Shawky E, Nahar L, Nassief S M, et al. Dairy products authentication with biomarkers: A comprehensive critical review[J]. Trends in Food Science & Technology, 2024, 147: 104445.

度的预测[1]。Feng 等将拉曼光谱与轻量梯度提升机、支持向量机、随机森林和极限梯度提升相结合,研究表明,在单一算法条件下,乳制品品牌分类的准确率超过90%,当这些算法结合使用时,准确率可达99%[2]。Zhao 等运用激光诱导击穿光谱、傅立叶变换中红外光谱和拉曼光谱数据与偏最小二乘回归算法相结合,实现了婴儿配方奶粉中的钙含量分析,结果显示出使用激光诱导击穿光谱数据开发的模型在钙含量定量分析方面取得了最佳性能,而基于拉曼和傅立叶变换中红外光谱的融合模型获得了第二好的性能[3]。邵帅斌等针对乳粉中可能存在的小麦粉、植脂末、乳清粉、滑石粉、糊精、淀粉等掺杂物,研发了一种基于卷积神经网络和拉曼光谱结合的判别分析方法,在实验体系下识别准确率可达95.5%[4]。

此外,扰动信息融合也引起了研究人员的广泛关注,其是通过设计外加微小扰动,定向收集乳制品检测体系在扰动干预情况下的变化图谱,经融合变换得到高维图谱。Huang 等以温度为扰动构建了牛乳的高维红外光谱,结合偏最小二乘判别法建立了纯牛乳和掺假牛乳的判别模型[5]。Wu 等以尿素掺杂为外扰构建了四个不同品牌牛乳的高维近红外光谱,结合偏最小二乘判别分析实现了掺假判别[6]。Al-Lafi 等以三聚氰胺为外扰构建了高维傅立叶变换红外光谱,开展了基于高维谱的掺假物定量分析[7]。扰动信息融合优势在于所构建的高维图谱综合利用了乳制品系统变化过程的多谱源数据与谱峰相关性信息,很好地刻画了乳制品体系在扰动条件下的动态变化趋势,且有望凸显乳制品常规图谱中的覆盖峰、弱

---

[1] Huang W, Fan D S, Li W F, et al. Rapid evaluation of milk acidity and identification of milk adulteration by Raman spectroscopy combined with chemometrics analysis[J]. Vibrational Spectroscopy, 2022, 123:103440.

[2] Feng Z K, Liu D, Gu J Y, et al. Raman spectroscopy and fusion machine learning algorithm: A novel approach to identify dairy fraud[J]. Journal of Food Composition and Analysis, 2024, 129:106090.

[3] Zhao M, Markiewicz-Keszycka M, Beattie R J, et al. Quantification of calcium in infant formula using laser-induced breakdown spectroscopy (LIBS), Fourier transform mid-infrared (FT-IR) and Raman spectroscopy combined with chemometrics including data fusion[J]. Food Chemistry, 2020, 320:126639.

[4] 邵帅斌,刘美含,石宇晴,等.基于卷积神经网络的乳粉掺杂物拉曼光谱分类方法[J].食品科学,2022,43(14):296-301.

[5] Huang M Y, Yang R J, Zheng Z Y, et al. Discrimination of adulterated milk using temperature-perturbed two-dimensional infrared correlation spectroscopy and multivariate analysis[J]. Spectrochimica Acta Part A, Molecular and Biomolecular Spectroscopy, 2022, 278:121342.

[6] Wu H Y, Yang R J, Wei Y, et al. Influence of brands on a discrimination model for adulterated milk based on asynchronous two-dimensional correlation spectroscopy slice spectra[J]. Spectrochimica Acta Part A, Molecular and Biomolecular Spectroscopy, 2022, 271:120958.

[7] Al-Lafi A G, AL-Naser I. Application of 2D-COS-FTIR spectroscopic analysis to milk powder adulteration: Detection of melamine[J]. Journal of Food Composition and Analysis, 2022, 113:104720.

峰、偏移峰等微小关键谱峰的变化规律,而这常常在原始图谱中是无法观测或易被忽略的。

1.3 基于拉曼光谱的乳制品相关研究进展

拉曼散射现象是印度科学家 C V Raman 在 1928 年首次发现的,由此获得了 1930 年的诺贝尔物理学奖。拉曼光谱是可以反映乳制品分子结构信息的一种表征技术手段,其基本原理可以做如下介绍[1][2]:当一束光照射到样品时,光子和物质分子会发生碰撞产生散射光,这时如果发生的只有光的方向改变,而没有能量交换,即光的频率不变的弹性碰撞,即被称为瑞利散射;与此相反,如果发生的既有光的方向改变也有能量交换,即光的频率也发生改变的非弹性碰撞,则被称为拉曼散射,对应的谱线图为拉曼光谱图。非弹性碰撞又可分为两种情况,如图 1-1 所示,一是产生斯托克斯(Stokes)线,指的是如果受激分子从基态振动能级跃迁到受激虚态的分子不返回基态,而返回到基态的高位能级,即分子保留一部分能量,此时散射光子的能量为 $h\upsilon - \Delta E$,式中,$h\upsilon$ 为入射光能量,ΔE 为基态电子能级第一振动激发态的能量,由此产生的拉曼线又被称为斯托克斯线。二是产生反斯托克斯线,指的是如果处于基态高位振动能级的分子跃迁到受激虚态后,再返回到基态振动能级,此时散射光子的能量则为 $h\upsilon + \Delta E$,产生的拉曼线即被称为反斯托克斯线。常温下处于基态的分子占绝大多数,故而斯托克斯线强于反斯托克斯线。在拉曼散射中,散射线频率和入射光频率存在一个频率差 $\Delta \upsilon = \Delta E/h$,这个频率差即为拉曼位移。在拉曼光谱图中,横坐标为拉曼位移,用波数表示;纵坐标为谱带强度。

拉曼光谱仪是采集拉曼光谱数据的检测设备,主要由光源、单色器、检测器等组成。由于拉曼散射现象较弱,为获取有效信号,目前常采用激光作为光源。图 1-2 展示了一种便携式拉曼光谱仪,将待测样品置于载物台或样品池,使用激光探头可以照射样品,单色器滤除杂散光,检测器收集得到拉曼信号。图 1-3 显示了乳制品置于激光探头下的测试示意图。

---

[1] 陈浩,汪圣尧.仪器分析[M].4 版.北京:科学出版社,2022:36-61.
[2] 武汉大学.分析化学:下册[M].5 版.北京:高等教育出版社,2007:291-302.

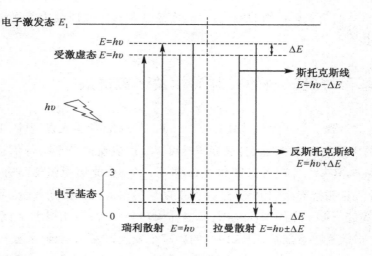

图 1-1 拉曼散射原理图

图 1-2 便携式拉曼光谱仪

图 1-3 基于拉曼光谱仪的乳制品检测示意图

拉曼光谱作为一种重要的乳制品检测表征手段,是一种散射光谱,可很好地表征样品分子振动信号,并具有样品用量少、采集速度快、可无损检测、仪器便携化等多种优点。由于水的散射截面较小,对拉曼光谱信号影响较小,含水样品亦可直接进行拉曼光谱测量,因此,拉曼光谱成为乳制品质量分析领域的重要数据基础。Hussain Khan 等考察了拉曼光谱法测量生乳中宏观成分(脂肪、蛋白质和乳糖)的适用性,研究建立了偏最小二乘回归模型预测生乳中脂肪、蛋白质和乳糖,预测均方根误差为 0.15、0.11 和 0.04,预测决定系数为 0.96、0.89 和 0.89,预测误差与偏差之比为 8.16、3.16 和 2.89[①]。Zhang 等针对乳制品非法添加剂硫氰酸钠,设计了一种表面增强拉曼光谱检测法,实现了牛奶中硫氰酸钠的高通量检测分析,检测限可达 5×10^{-7} g/mL[②]。Yang 等采集分析了液态奶中三聚氰胺的表面增强拉曼光谱,并利用拉曼特征峰和拉曼强度对三聚氰胺进行了定性鉴定和半定量分析,研发出了一种简单高效的三聚氰胺快速筛选方法,检测限为 0.25 mg/kg[③]。Nedeljkovic 等进行了不同乳脂/植物脂肪比例的奶油及其与葵花子油、椰子油和棕榈油类似物的拉曼光谱分析,结合化学计量学方法实现了乳制品奶油和奶油状类似物的分类[④]。

综上所述,可以获知现有报道从多个方面围绕乳制品开展着持续深入的研究工作,建立了包括拉曼光谱在内的多种检测表征技术,开发出针对多个目标物质的多种分析方法,实现了多维检测表征数据与智能识别算法的结合探索,极大地提升了人们对于乳制品质量要素的认知程度,积累了丰富的乳制品质量评价分析方法研发经验。本书在此背景下,聚焦拉曼光谱与乳制品质量判别技术领域,结合多年研究工作积累,通过实验案例研究,尝试从光谱数据预处理、特征提取、智能判别等多个视角进行了系统小结,以期为乳制品质量科学管控提供技术参考。

---

① Hussain Khan H M, McCarthy U, Esmonde-White K, et al. Potential of Raman spectroscopy for in-line measurement of raw milk composition[J]. Food Control, 2023, 152: 109862.
② Zhang Z Y, Liu J, Wang H Y. Microchip-based surface enhanced Raman spectroscopy for the determination of sodium thiocyanate in milk[J]. Analytical Letters, 2015, 48(12): 1930 - 1940.
③ Yang Q L, Deng X J, Niu B, et al. Qualitative and semi-quantitative analysis of melamine in liquid milk based on surface-enhanced Raman spectroscopy[J]. Spectrochimica Acta Part A, Molecular and Biomolecular Spectroscopy, 2023, 303: 123143.
④ Nedeljkovic A, Tomasevic I, Miocinovic J, et al. Feasibility of discrimination of dairy creams and cream-like analogues using Raman spectroscopy and chemometric analysis[J]. Food Chemistry, 2017, 232: 487 - 492.

第 2 章 乳制品拉曼光谱数据的预处理研究

在乳制品质量判别研究过程中,拉曼光谱可以表征其组分特性。不过,研究也发现拉曼光谱数据采集会受到环境条件、检测仪器等多种因素的影响,实际获取的拉曼光谱可能存在噪声、受到实验条件的影响等问题,为此,需要开展光谱数据预处理实验研究,了解其变化规律,以期建立科学的处置方法。

2.1 乳粉拉曼光谱表征数据的标准化与降噪处理研究

乳制品是大众日常消费品的重要组成部分,自 2008 年三聚氰胺事件爆发以来,乳制品的质量安全问题一直是监管部门和普通民众关注的热点问题之一。现有的管控思路主要是遵循国家标准,产品标准有如《食品安全国家标准 乳粉》(GB 19644—2010),检测标准有如《食品安全国家标准 乳和乳制品杂质度的测定》(GB 5413.30—2016)等,起到了控制乳制品质量安全风险的重要作用,积累了大量检测数据。不过,随着新需求的出现,该方法存在进一步改进的空间。如偶见媒体报道的假冒乳粉事件,以次充好,牟取暴利,是一种新型的乳粉质量安全问题,而传统的成分检测手段难以及时发现[1][2]。此外,工厂乳制品生产涉及多个加工环节,加强过程能力控制,可望改变传统事后检测的管控策略,变被动为主动,在质量变异时及时采取纠偏措施,将有望减少质量损失。为此,本研究小组在研究过程中提出了谱图数据支持的乳制品质量控制新策略,基本思路是收集乳制品的谱图数据,在乳制品产品仅受正常波动因素影响时,谱图数据间相似度很高;而出现异常因素时,数据间相似度将有明显下降或有规律降低,结合质量波动控

---

[1] 徐建华. 我国乳业将迎来更严监管:专家解读上海假奶粉事件[N]. 中国质量报,2016-04-08(2).
[2] 史若天. 探析公共食品安全事件中政府的舆论引导策略:以 2016 年上海"假奶粉"事件为例[J]. 新闻研究导刊,2016,7(12):334.

制图或模式识别算法可进行质量预警[1][2]。拉曼光谱是一种表征样品分子振动信号的光谱检测技术,具有样品用量少、采集速度快、可无损检测、仪器便携化等优点,在行政现场执法、实验室检测、工厂在线检测等领域具有广阔的应用前景。拉曼光谱用于乳制品表征研究的相关报道近年来逐年增多,是乳制品表征分析的热点手段之一[3][4]。

发展基于拉曼光谱的乳制品质量分析与质量预警研究,关键因素是采集的拉曼光谱数据,数据的瑕疵将导致判别模型结果不可靠。因此,本节研究了乳制品拉曼光谱表征数据的标准化处理,提出了针对不同噪声水平时数据规范化处理评价的适用方法,研究为破解时下检测数据"信息孤岛"问题提供了技术参考,为数据共享研究提供了参考思路[5][6]。

2.1.1 实验部分

1. 材料

实验用乳粉购置于南京苏果超市,其中,贝因美乳粉标记为品牌 P1,合生元乳粉标记为品牌 P2。

2. 仪器与设备

激光拉曼光谱仪,光谱仪型号:Prott-ezRaman-D3,厂家:美国恩威光电公司(Enwave Optronics),激光波长为 785 nm,激光最大功率约为 450 mW,电荷耦合器件检测器温度为 -85 ℃。96 孔板:美国康宁公司(Corning Incorporated)。

3. 拉曼光谱谱图采集方法

取适量乳粉粉末置于 96 孔板的独立小孔内,保持小孔恰好处于充满状态。而后,使用激光拉曼光谱仪直接照射样品,进行测试,收集测试信号即得到乳粉的

---

[1] 张正勇,沙敏,刘军,等. 基于高通量拉曼光谱的奶粉鉴别技术研究[J]. 中国乳品工业,2017,45(6):49-51.

[2] 张正勇,沙敏,冯楠,等. 基于统计过程控制的液态奶脱脂工序评价分析[J]. 食品安全导刊,2017(28):66-69.

[3] Zhang Z Y, Liu J, Wang H Y. Microchip-based surface enhanced Raman spectroscopy for the determination of sodium thiocyanate in milk[J]. Analytical Letters, 2015, 48(12): 1930-1940.

[4] Nieuwoudt M K, Holroyd S E, McGoverin C M, et al. Rapid, sensitive, and reproducible screening of liquid milk for adulterants using a portable Raman spectrometer and a simple, optimized sample well[J]. Journal of Dairy Science, 2016, 99(10): 7821-7831.

[5] 张正勇,李丽萍,岳彤彤,等. 乳粉拉曼光谱表征数据的标准化与降噪处理研究[J]. 粮食科技与经济,2018,43(6):57-61.

[6] 王海燕,等. 食药质量安全检测技术研究[M]. 北京:科学出版社,2023:1-215.

拉曼光谱图。激光功率分别采用～250 mW，～350 mW，～450 mW，照射时间为 50 s，光谱范围为 250～2 339 cm^{-1}，光谱分辨率为 1 cm^{-1}。

4. 数据分析方法

（1）平均值标准化处理：计算公式如式（2.1）所示，式中，x_i 表示的是拉曼光谱的强度值，y_i 表示的是标准化后的拉曼光谱的强度值，$\bar{x}$ 表示的是 x_i 的平均值，计算公式如 $\bar{x} = \dfrac{\sum\limits_{i=1}^{n} x_i}{n}$，$s$ 表示的是标准偏差，计算公式如 $s = \sqrt{\dfrac{\sum\limits_{i=1}^{n}(x_i - \bar{x})^2}{n-1}}$。

$$y_i = \frac{x_i - \bar{x}}{s} \tag{2.1}$$

（2）极大值标准化处理：计算公式如式（2.2）所示，式中，x_i 表示的是拉曼光谱的强度值，$\max(x_i)$ 表示的是拉曼光谱中谱峰最大强度值，本实验中选取 1 450 cm^{-1} 的峰值为最大值，z_i 表示的是标准化后的拉曼光谱的强度值。

$$z_i = \frac{x_i}{\max(x_i)} \tag{2.2}$$

（3）小波降噪，使用小波软阈值函数进行光谱降噪处理，计算公式如式（2.3）所示，式中，x 为光谱信号值，T 为阈值，s_T 为重构信号。

$$s_T = \begin{cases} \text{sign}(x)(|x| - T), & |x| > T \\ 0, & |x| \leqslant T \end{cases} \tag{2.3}$$

（4）相关系数计算：计算公式如式（2.4）所示，式中，R 表示两样品间的相关系数值，a_i 和 b_i 分别表示的是两样本在同一波长处的拉曼光谱强度值；$\bar{a}$ 和 $\bar{b}$ 分别表示两样本拉曼光谱强度的平均值。相关系数值愈接近于 1，说明两者愈正相关，相关系数值愈接近于 0，说明两者愈不相关。数据运算分析平台：MATLAB R2016a。

$$R = \frac{\sum\limits_{i=1}^{n}(a_i - \bar{a})(b_i - \bar{b})}{\sqrt{\sum\limits_{i=1}^{n}(a_i - \bar{a})^2 \sum\limits_{i=1}^{n}(b_i - \bar{b})^2}} \tag{2.4}$$

2.1.2 结果与讨论

1. 乳粉拉曼光谱解析

实验首先分别采集了在激光功率～250 mW，～350 mW，～450 mW 条件下，国内 P1 品牌乳粉的拉曼光谱数据，如图 2-1 所示。谱图表征了乳粉丰富的组分

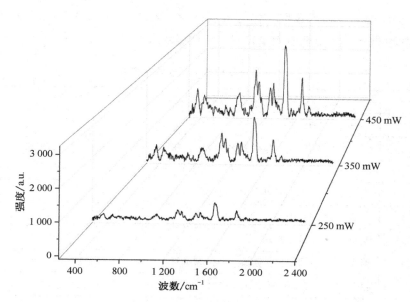

图 2-1 P1 品牌乳粉不同激光功率条件下的拉曼光谱图

信息,结合已有文献报道,可进行谱峰归属分析[1][2][3][4],在图示乳粉的拉曼光谱图中谱峰 1 746 cm^{-1} 主要来自脂肪有关的酯基 C═O 伸缩振动,1 667 cm^{-1} 主要来自蛋白质的酰胺 I 键的 C═O 伸缩振动和不饱和脂肪酸的 C═C 伸缩振动,1 450 cm^{-1} 主要来自糖类和脂肪的 CH_2 变形振动,1 009 cm^{-1} 主要来自蛋白质苯丙氨酸的苯环振动,详细的拉曼光谱的谱峰解析结果参见表 2-1。从拉曼光谱解析结果可以看出,拉曼光谱可以表征显示出乳粉的糖类、脂肪和蛋白质等主要化学组分信息,其峰位置、峰高、峰面积蕴含着乳粉组成分子的种类、含量信息。在过往研究中,发现在生产工序稳定、原料来源一致等质量要素可控条件下,乳粉的拉曼光谱表征数据将表现出较高的一致性,而质量变异因素出现时拉曼光谱将出

---

[1] Rodrigues Júnior P H, de Sá Oliveira K, deAlmeida C E R, et al. FT-Raman and chemometric tools for rapid determination of quality parameters in milk powder: Classification of samples for the presence of lactose and fraud detection by addition of maltodextrin[J]. Food Chemistry, 2016, 196: 584-588.

[2] Zhang Z Y, Sha M, Wang H Y. Laser perturbation two-dimensional correlation Raman spectroscopy for quality control of bovine colostrum products[J]. Journal of Raman Spectroscopy, 2017, 48(8): 1111-1115.

[3] Almeida M R, de S Oliveira K, Stephani R, et al. Fourier-transform Raman analysis of milk powder: A potential method for rapid quality screening[J]. Journal of Raman Spectroscopy, 2011, 42(7): 1548-1552.

[4] 王海燕,宋超,刘军,等. 基于拉曼光谱—模式识别方法对奶粉进行真伪鉴别和掺伪分析[J]. 光谱学与光谱分析,2017,37(1):124-128.

现谱峰变化，据此可提示我们一定的质量风险。因此，依据乳粉的拉曼光谱表征数据有望成为构建新型乳粉质量控制体系的技术方案。

表2-1 乳粉拉曼光谱峰归属解析表

波数/cm^{-1}	归属
1 746	C＝O伸缩振动，主要可能源自脂肪有关的酯基
1 667	C＝O伸缩振动和C＝C伸缩振动，其中C＝O伸缩振动可能主要源自蛋白质的酰胺Ⅰ键，C＝C伸缩振动主要源自不饱和脂肪酸
1 450	CH$_2$变形振动，可能主要源自糖类和脂肪分子
1 312	脂肪酸的CH$_2$扭曲振动
1 270	糖类的CH$_2$扭曲振动
1 130	糖类的C—C伸缩振动、C—O伸缩振动以及C—O—H变形振动
1 091	糖类的C—C伸缩振动、C—O伸缩振动以及C—O—H变形振动
1 009	蛋白质苯丙氨酸的苯环振动
950	糖类的C—O—C变形振动、C—O—H变形振动和C—O伸缩振动
885	糖类的C—C—H变形振动和C—O—C变形振动
767	糖类的C—C—O变形振动
705	蛋白质的C—S伸缩振动
627	糖类的C—C—O变形振动
592	糖类的C—C—C变形振动、C—O扭曲振动
516	葡萄糖
447	糖类的C—C—C变形振动、C—O扭曲振动
360	乳糖

2. 乳粉拉曼光谱标准化处理分析

数据驱动型的乳制品质量控制新技术，关键要素是采集的拉曼光谱数据，因此有必要建立标准化处理流程[①②]。如图2-1所示，本实验尝试选择了三种不同的激光功率分别采集P1品牌乳粉拉曼光谱数据，图示可以看出尽管测试对象一

---

① Grelet C, Pierna J A F, Dardenne P, et al. Standardization of milk mid-infrared spectrometers for the transfer and use of multiple models[J]. Journal of Dairy Science, 2017, 100(10): 7910-7921.

② 孙雪杉,杨仁杰,杨延荣,等.不同预处理方法对二维相关谱的影响研究Ⅰ:标准化方法[J].天津农学院学报,2015,22(4):13-16,20.

致,但各实验条件下拉曼光谱谱图有一定差异,具体表现为不同激光功率条件下,各谱图的峰强度不一致,且峰强度随着激光强度的增加而增加。实验选择了平均值标准化处理和极大值标准化处理方法对实验数据进行了标准化处理,图 2-2 显示了平均值标准化处理的乳粉拉曼光谱谱图,可以明显看出经过处理后,不同激光功率获得的乳粉拉曼光谱谱峰强度差异得以消除。极大值标准化处理方法以乳粉拉曼光谱最高峰 1 450 cm^{-1} 处的强度值为基准,按照前述公式进行了计算,结果如图 2-3 所示,显示在纵坐标上有一定不同,整体图示与图 2-2 类似。实验进一步以三种不同激光功率采集的乳粉拉曼光谱数据的平均值为真值,以相关系数为相似度评价指标,分别计算三种不同激光功率采集的乳粉拉曼光谱数据与平均值间的相关系数,量化计算比较未做标准化处理以及两种不同标准化处理方法的结果差异情况,如图 2-4 所示。结果显示,未做标准化处理其相关系数为 0.986 4±0.012 1,平均值标准化处理其相关系数为 0.988 6±0.003 9,极大值标准化处理其相关系数为 0.988 6±0.004 4。其中,0.986 4,0.988 6 是均值,0.012 1,0.003 9,0.004 4 是标准偏差。不难看出,标准化处理后,去除了量纲,三种不同激光功率采集的乳粉拉曼光谱数据与样品均值的相关性有所提高,分散性有所改善,且平均值标准化处理效果最好。同时,研究发现图 2-4 所示相关系数结果虽接近与 1,但均未达到 1,可能是由于以下原因:光谱采集过程中随机噪声的存在,图 2-2 所示也可以看出,250 mW 条件下采集时噪声较大。

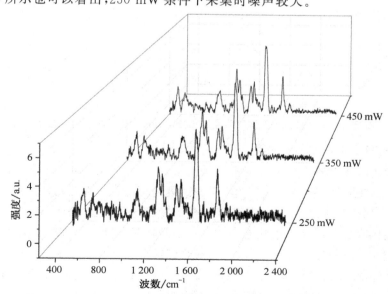

图 2-2 P1 品牌乳粉不同激光功率条件下的拉曼光谱平均值标准化处理结果图

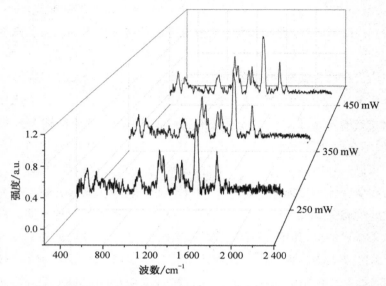

图 2-3 P1 品牌乳粉不同激光功率条件下的拉曼光谱极大值标准化处理结果图

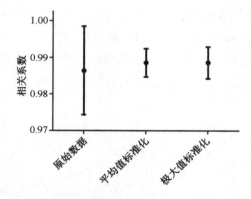

图 2-4 P1 品牌乳粉不同标准化方法处理后的相关系数结果图

3. 乳粉拉曼光谱小波降噪分析

实验进一步对噪声影响及降噪处理进行了分析,首先计算了平均值标准化处理后 P1 品牌乳粉三种不同激光功率采集的乳粉拉曼光谱数据与样品均值的相关系数,结果如表 2-2 所示,达到 0.98 以上。随机选择了市售 P2 品牌乳粉,按照同样的操作收集其拉曼光谱数据,并与 P1 品牌样品光谱均值进行相关系数运算,结果如表 2-2 所示,均在 0.98 以下。据此可以区分品牌 P1 和 P2。前述观察图 2-2 可知,噪声影响在 250 mW 条件下最大,350 mW 次之,450 mW 最小,因此,为保证判别模型的准确性,降噪处理是非常有必要的。小波降噪的基本思路是采用低通滤波器减少噪声强度,去除原始谱图中的棘波,增强信噪比[1][2]。本节应用小波软阈值进行降噪处理,基本步骤是经过小波变换后真实信号与噪声的统计特性不同,在小波分解后的各层系数中,进行阈值处理,而后再反变换重构,小波软阈值可以有效避免间断,使得重构的信号更加光滑。常用的小波基有 Daubechies(dbN)、Symlets(symN)、Coiflet(coifN)、Biorthogonal(biorNr. Nd)四种小波基,实验设定阈值选定方法 rigrsure,分解层数设定为三层,考察了不同小波基条件下 P1 品牌和 P2 品牌乳粉三种不同激光功率采集的乳粉拉曼光谱数据与 P1 品牌样品均值的相关系数,结果见表 2-3。由于噪声具有随机性,会一定程度降低样品间相关系数,经过降噪处理后,相关系数值得以提高。表 2-3 显示,P1 品牌乳粉三种不同激光功率与 P1 品牌样品均值的相关系数值经过小波处理几乎都达到了 0.99 以上,其中,db1、sym1、bior1.1 处理后相关系数值达到最大值。与此同时,P2 品牌乳粉三种不同激光功率与 P1 品牌样品均值的相关系数值经过小波处理后也有不同程度的提高,但均仍低于 0.99。图 2-5 显示了小波降噪后的效果图,与图 2-2 未进行小波处理结果相比,可以明显看出,谱线变得更加光滑,噪声明显减弱。

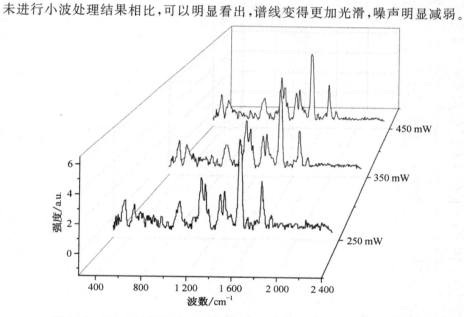

图 2-5　db1 小波处理后 P1 品牌乳粉不同激光功率条件下的拉曼光谱图

---

[1]　Ehrentreich F, Sümmchen L. Spike removal and denoising of Raman spectra by wavelet transform methods[J]. Analytical Chemistry, 2001, 73(17): 4364-4373.

[2]　Avohou T H, Sacré P Y, Hubert P, et al. Interpretable one-class classification of Raman spectra using prediction bands estimated by wavelet regression[J]. Analytical Chemistry, 2022, 94(10): 4183-4191.

表 2-2 平均值标准化后 P1 品牌乳粉与 P2 品牌乳粉相关系数计算结果

相关系数	P1 品牌 (~250 mW)	P1 品牌 (~350 mW)	P1 品牌 (~450 mW)	P2 品牌 (~250 mW)	P2 品牌 (~350 mW)	P2 品牌 (~450 mW)
P1 品牌（均值）	0.983 1	0.991 2	0.991 5	0.953 5	0.974 7	0.979 6

表 2-3 不同小波降噪处理、均值标准化后 P1 品牌乳粉、P2 品牌乳粉与 P1 品牌（均值）相关系数计算结果

降噪方法	P1 品牌 (~250 mW)	P1 品牌 (~350 mW)	P1 品牌 (~450 mW)	P2 品牌 (~250 mW)	P2 品牌 (~350 mW)	P2 品牌 (~450 mW)
db1	0.991 1	0.995 6	0.995 2	0.974 2	0.983 1	0.984 9
db2	0.990 6	0.995 0	0.995 1	0.977 0	0.983 0	0.985 4
db3	0.991 3	0.995 5	0.995 4	0.975 7	0.983 7	0.985 3
db4	0.991 0	0.995 2	0.995 1	0.973 3	0.982 9	0.985 3
db5	0.990 6	0.995 2	0.995 1	0.975 8	0.983 4	0.984 9
sym1	0.991 1	0.995 6	0.995 2	0.974 2	0.983 1	0.984 9
sym2	0.990 6	0.995 0	0.995 1	0.977 0	0.983 0	0.985 4
sym3	0.991 3	0.995 5	0.995 4	0.975 7	0.983 7	0.985 3
sym4	0.990 8	0.995 4	0.995 2	0.975 1	0.983 4	0.985 3
sym5	0.990 5	0.995 1	0.994 7	0.975 0	0.983 6	0.985 0
coif1	0.990 7	0.995 4	0.995 0	0.976 4	0.983 8	0.985 0
coif2	0.990 9	0.995 2	0.995 1	0.974 4	0.982 3	0.985 0
coif3	0.990 4	0.995 0	0.995 0	0.976 1	0.983 5	0.984 7
coif4	0.990 7	0.995 0	0.995 0	0.974 7	0.982 5	0.984 5
coif5	0.990 4	0.995 0	0.995 0	0.975 1	0.983 3	0.984 6
bior1.1	0.991 1	0.995 6	0.995 2	0.974 2	0.983 1	0.984 9
bior1.3	0.990 2	0.994 7	0.995 1	0.971 9	0.981 3	0.983 9
bior1.5	0.989 7	0.994 5	0.994 3	0.973 2	0.981 6	0.984 2
bior2.2	0.990 5	0.995 3	0.994 9	0.976 2	0.984 3	0.985 2
bior2.4	0.990 9	0.995 3	0.995 1	0.975 0	0.982 8	0.984 9

2.1.3 小结

拉曼光谱可以表征乳制品丰富的质量特征信息,通过拉曼光谱数据分析与处理可用以乳制品的品质控制,其中质量评价的关键因素是表征数据的规范化处理。本节研究探讨了三种不同激光功率条件下,乳粉拉曼光谱数据的标准化与降噪处理。通过平均值标准化处理,可以有效消除量纲影响,减少不同条件下采集数据的分散性。通过小波降噪处理,可以有效减少拉曼光谱信号采集过程中引入的随机噪声,提高信噪比,凸显有效信号。本节研究揭示出平均值标准化处理、db1(sym1、bior1.1)小波降噪手段的使用,适用于乳粉拉曼光谱表征数据的规范化前处理,有利于保证后续判别模型的准确性。

2.2 基于小波变换的乳制品智能鉴别技术优化研究

在全面建设社会主义现代化国家的新征程中,人们日益增长的高品质食品需求与潜在的制假售假间的矛盾成为食品质量安全管理领域关注的核心问题之一,鉴别技术是解决这一问题的关键支撑。乳制品作为食品的重要组成部分,其质量安全评价一直是监管部门和消费者关注的重点[1][2]。现有的鉴别方法可以主要分为以下四种类型:一是感官检验,主要是依靠品鉴专家的感觉器官进行判断,具有一定的主观性。二是理化检测法,主要依据一些物理特性值如旋光性[3],或者一些特征化学成分如氨基酸、同位素进行鉴别[4][5][6][7]。三是生化检测法,主要是依

---

[1] 李思维,孙树垒,张正勇.大学生液态奶消费行为研究:以南京市仙林大学城为例[J].粮食科技与经济,2019,44(6):104-108.

[2] 姜冰,李翠霞.基于宏观数据的乳制品质量安全事件的影响及归因分析[J].农业现代化研究,2016,37(1):64-70.

[3] 郑向华,杨自洁,龚吉军,等.基于旋光法的原料乳中乳糖掺伪鉴别技术研究[J].食品工业科技,2015,36(6):75-77,89.

[4] 易冰清,郭秀秀,颜治,等.乳制品掺假现状与稳定同位素鉴别技术研究进展[J].同位素,2020,33(5):293-303.

[5] 赵超敏,王敏,张润何,等.碳氮稳定同位素鉴别有机奶粉[J].现代食品科技,2018,34(12):211-215.

[6] 宋蓓,宋桂雪,宋薇.羊乳制品中牛乳源成分的鉴别检测技术[J].食品科学技术学报,2016,34(6):69-74.

[7] 钱宇,汪慧超,吴林昊,等.超高效液相色谱检测乳粉复原乳中的氨基酸[J].食品与机械,2016,32(7):56-60.

据乳品基因序列进行判别[1][2]。四是计算机智能鉴别算法与谱图数据相结合进行模式识别[3][4][5]，较之前的三种方法，计算机辅助鉴别技术具有运算速度快、结果评价客观等优势，成为鉴别技术研发的热点。

谱图数据可以表征乳制品的化学质量特性，并作为智能鉴别算法的数据输入，经过判别函数运算构建鉴别模型，用于未知样品类别归属测算，可见谱图数据作为鉴别模型的重要组成部分，影响着模型的识别准确率。在2.1节研究中，基于小波降噪处理可以实现拉曼光谱数据的平滑，达到凸显信号抑制噪声的滤波效果。在本节中，通过采集不同品牌乳酪制品的拉曼光谱数据，结合 k 近邻鉴别算法，进一步考察了小波软阈值降噪、小波信号增强、归一化处理、数据融合等多种数据预处理技术与判别算法相结合的情况，以期建立一个新型的优化识别流程，为挖掘谱图数据信息、提高乳制品鉴别模型适应性提供技术参考[6]。

2.2.1 实验部分

1. 材料

实验用乳酪制品均购置于南京苏果超市，原味口味，选取3个品牌，伊利乳酪标记为品牌P1，妙可蓝多乳酪标记为品牌P2，夏洛克乳酪标记为品牌P3，每个品牌25个样品，共计75个样品。

2. 仪器与设备

激光拉曼光谱仪，光谱仪型号：Prott-ezRaman-D3，厂家：美国恩威光电公司（Enwave Optronics），激光波长为785 nm，激光最大功率约为450 mW，电荷耦合器件检测器，温度为-85℃，样品采集照射时间为80 s，扫描次数为1次，光谱波数收集范围为250～2 000 cm^{-1}，光谱分辨率为1 cm^{-1}。上样载体为96孔板，厂

---

[1] 王之莹,李婷婷,于文杰,等.一种适用于乳制品基因组DNA快速提取方法的研究[J].食品安全质量检测学报,2020,11(1):134-139.

[2] 李富威,张舒亚,曾庆坤,等.乳制品中水牛乳成分的实时荧光PCR检测技术[J].农业生物技术学报,2013,21(2):247-252.

[3] 王梓笛,李双妹,尹延东,等.基于支持向量机算法的乳制品分类识别技术研究[J].粮食科技与经济,2020,45(3):104-107.

[4] 黄宝莹,佘之蕴,王文敏,等.近红外光谱技术在乳制品快速检测中的应用研究进展[J].中国酿造,2020,39(7):16-19.

[5] 荣菡,甘露菁.基于近红外光谱的自组织映射神经网络快速鉴别牛乳与掺假乳[J].食品工业,2019,40(8):188-191.

[6] 黄文萍,赵依琳,杨如玲,等.基于小波变换的乳酪制品智能鉴别技术[J].粮食科技与经济,2021,46(1):127-130.

家:美国康宁公司(Corning Incorporated)。

3. 数据采集与处理

取一定量的乳酪制品置于96孔板的各独立小孔内,使得小孔恰好被样品充满。而后,将激光拉曼光谱仪的探头固定于小孔上方,恰好可以直射样品,此时,测试收集信号即为乳酪制品的拉曼光谱数据。

测试获得样品谱图数据后,使用光谱仪自带的 SLSR Reader V8.3.9 软件进行基线校正,校正后的谱图数据归一化处理使用 mapminmax 函数,归一化至[0,1]区间。小波软阈值降噪使用 wden 函数,小波增强使用 wavedec 函数进行谱图数据小波分解,而后使用 waverec 函数进行小波重构。小波变换处理、归一化、k 近邻算法的运算平台采用 MATLAB 软件:美国 MathWorks 公司,版本为 2016a。

2.2.2 结果与讨论

1. 乳酪制品的拉曼光谱表征分析

拉曼光谱是一种利用光子与物质相互作用发生散射效应,主要反映物质分子振动信息的指纹图谱。图 2-6 所示是实验采集的三种不同品牌乳酪制品的拉曼光谱图,分别标记为品牌 P1(a),品牌 P2(b)和品牌 P3(c),乳酪制品均为乳黄色含水分的黏稠状样品。拉曼光谱可以直接上样测试,无需样品前处理,且由于水分子的拉曼散射截面较小,无明显拉曼信号也不影响拉曼测试。从图中可以看出,3 个品牌乳酪制品的拉曼光谱十分相似,在同一个波数附近都出现了波峰,起伏程度也十分接近,结合已有相关文献报道,可知该类波峰的产生主要源于乳制品中糖类、脂肪、蛋白质分子相关的变形振动、伸缩振动与扭曲振动等[1][2][3]。此外,这三类乳酪制品拉曼光谱在 800 cm^{-1} 至 1 800 cm^{-1} 区间振动较为剧烈,出峰明显。与糖类有关的拉曼谱峰有 850 cm^{-1},源于 C—C—H 变形振动和 C—O—C 变形振动;940 cm^{-1} 源于 C—O—C 变形振动、C—O—H 变形振动以及 C—O 伸

---

[1] Rodrigues Júnior P H, de Sá Oliveira K, deAlmeida C E R, et al. FT-Raman and chemometric tools for rapid determination of quality parameters in milk powder: Classification of samples for the presence of lactose and fraud detection by addition of maltodextrin[J]. Food Chemistry, 2016, 196: 584-588.

[2] 张正勇,沙敏,刘军,等. 基于高通量拉曼光谱的奶粉鉴别技术研究[J]. 中国乳品工业,2017,45(6):49-51.

[3] Almeida M R, de S Oliveira K, Stephani R, et al. Fourier-transform Raman analysis of milk powder: A potential method for rapid quality screening[J]. Journal of Raman Spectroscopy, 2011, 42(7): 1548-1552.

缩振动；1 079 cm^{-1}、1 145 cm^{-1}源于C—O—H变形振动、C—O伸缩振动以及C—C伸缩振动；与脂肪有关的峰有1 314 cm^{-1}，源于脂肪酸的CH_2扭曲振动；1 760 cm^{-1}源于C=O伸缩振动；与蛋白质有关的峰有1 019 cm^{-1}，源于苯丙氨酸的苯环振动即环内C—C对称伸缩振动；最高峰1 457 cm^{-1}则是源自糖类和脂肪分子CH_2变形振动；1 670 cm^{-1}是源自蛋白质的酰胺Ⅰ键C=O伸缩振动和不饱和脂肪酸C=C伸缩振动。可见，拉曼光谱表征出丰富的乳酪制品分子组分信息，但由于谱图信号较为相似，仅凭人眼难以实现高效分类鉴别，需要借助计算机识别算法开展进一步判别分析。

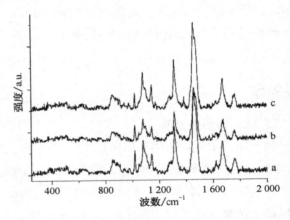

图2-6　不同品牌P1(a)、P2(b)和P3(c)乳酪制品的拉曼光谱图

2. 基于乳酪制品拉曼光谱的小波降噪分析

将采集获得的乳酪制品拉曼光谱数据导入k近邻算法，判别条件为马氏距离，$k=1$，随机选择80%的样品数据为训练集，余下20%的样品数据为测试集，重复随机循环测试100次，计算得出测试平均值，记为识别率，结果为82.27%[①]。此结果表明，一方面实验样品的拉曼谱图较为相似，直接使用原始数据进行类别判别，识别率有限；另一方面拉曼光谱原始数据中含有随机噪声、冗余信息，制约了鉴别模型的准确性。实验进一步运用小波变换方法的多尺度、多分辨特性，首先进行了小波降噪处理，基本思路是将拉曼光谱信号进行小波分解，保留高于阈值的小波系数，滤除小于阈值的噪声系数，而后通过逆小波重构获得降噪后的谱图数据。实验采用wden函数，综合比较分析了小波基(wname)、分解尺度(n)、

---

① 张正勇，岳彤彤，马杰，等.基于拉曼光谱与k最近邻算法的酸奶鉴别[J].分析试验室，2019，38(5)：553-557.

阈值处理噪声水平(scal)、函数选择阈值使用方式(sorh)以及阈值选择标准(tptr)参数条件下的模型识别率。实验采用小波软阈值去噪法,因此在考虑阈值使用方式(sorh)时,设定 sorh=s,(注:sorh=s,为软阈值;sorh=h,为硬阈值)。设置 tptr 的变量为 4 个,分别为:rigrsure、heursure、sqtwolog 和 minimaxi;设置 scal 的变量为 3 个,分别为:one、sln 和 mln;设置分解尺度为 5 个,即 $n=1、2、3、4、5$,研究了 4 类小波中各 5 个常见小波基,分别为 Biorthogonal 小波系中 bior1.1、bior1.5、bior2.2、bior2.4 与 bior3.1,Coiflet 小波系中的 coif1、coif2、coif3、coif4 与 coif5,Daubechies 小波系中 db1、db2、db3、db4 和 db5,Symlets 小波系中 sym1、sym2、sym3、sym4 和 sym5。最终选择了 tptr=heursure,sorh=s,scal=mln,$n=4$,wname=coif1 为小波降噪条件,降噪后的乳酪制品拉曼光谱结果如图 2-7 所示,识别率为 86%。此结果表明,一方面运用合适的小波降噪方法,可以有效降低随机噪声,改善谱图质量;另一方面,也减少了噪声对模型识别效果的影响,在一定程度上提高了分类算法的准确率。

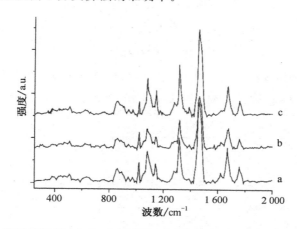

图 2-7 不同品牌 P1(a)、P2(b)和 P3(c)乳酪制品小波降噪后的拉曼光谱图

3. 基于乳酪制品拉曼光谱的小波增强、融合分析

小波降噪后分类算法识别率有所上升,提示我们谱图各波段对于识别结果的影响有所不同,进一步开展特征提取实验。首先采用小波增强方法,增强有效贡献信号,基本思路是采用小波函数对谱图数据进行分解,而后对分解系数进行选择性增强与削弱,再通过小波重构获取处理后的谱图。实验选用 sym5 小波函数对前述拉曼光谱数据进行了两层分解,对于>100 的小波分解系数赋予两倍增强,对于≤100 的小波分解系数赋予 0.5 倍削弱,随后,在此基础上进行谱图重构,结果如图 2-8 所示,识别率为 87.07%。此结果表明,小波增强法进一步凸显了特征波段,在一定程度上进一步提高了识别率。进一步将图 2-8 所示谱图波段,划分为 10 个特征波段区间,计算得到对应的识别率如表 2-4 所示。由此可以看出,各波段对识别率影响确有不同,可进一步开展融合分析。将 10 个特征波段区间融合,识别结果为 86.8%。而当 1 240~1 400 cm^{-1} 与 1 595~1 710 cm^{-1} 波段融合后,识别率可以达到 88.6%,与此同时,其运算时间也由全波段数据小波增强后的 1.15 s 减少到 0.6 s,提高了约 48% 的运算效率。

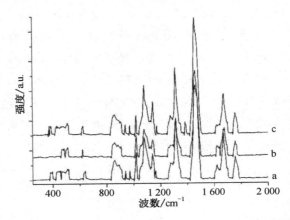

图 2-8　不同品牌 P1(a)、P2(b) 和 P3(c) 乳酪制品小波增强后的拉曼光谱图

表 2-4　基于乳酪制品拉曼光谱的特征波段及对应识别结果

波段区间/cm^{-1}	识别率/%	波段区间/cm^{-1}	识别率/%
280~560	48.93	1 035~1 185	60.85
595~670	39.40	1 240~1 400	85.13
820~925	51.53	1 415~1 515	65.73
925~990	62.47	1 595~1 710	70.80
1 000~1 035	51.33	1 730~1 790	56.67

4. 基于乳酪制品归一化拉曼光谱的小波增强、融合分析

实验进一步研究了归一化处理对模型识别效果的影响,通过归一化处理可以消除量纲的影响,减少数据的分散性,以提高识别效率[①]。k 近邻算法设置同前,乳酪制品拉曼光谱数据经 coif1 小波降噪及归一化处理后导入分类算法,实验数据归一化至[0,1]区间,识别率为 93.73%,识别率较前有较大提高。进一步进行

---

① 张正勇,李丽萍,岳彤彤,等.乳粉拉曼光谱表征数据的标准化与降噪处理研究[J].粮食科技与经济,2018,43(6):57-61.

小波增强,实验条件为选用 sym5 小波函数对拉曼光谱数据进行了两层分解,对于>0.2 的小波分解系数赋予 1 倍增强,对于≤0.2 小波分解系数赋予 0.5 倍削弱,随后,在此基础上进行谱图重构,识别结果为 94.4%。进一步将小波增强后的拉曼光谱谱图波段,划分为 10 个特征波段区间,计算得到对应的识别率如表 2-5 所示。结果也表明不同波段对识别率的影响不同,且经过归一化后,识别率最高波段为波段 1 595~1 710 cm^{-1},不同于未做归一化处理时的波段 1 240~1 400 cm^{-1}。进一步开展融合分析,将 10 个特征波段区间融合,识别结果为 92.6%。而当 1 415~1 515 cm^{-1} 与 1 595~1 710 cm^{-1} 波段融合后,识别率可以达到 95.4%,与此同时,其运算时间也同样由全波段数据小波增强归一化处理后的 1.15 s 减少到 0.6 s,提高了约 48% 的运算效率。

表 2-5　基于乳酪制品归一化拉曼光谱的特征波段及对应识别结果

波段区间/cm^{-1}	识别率/%	波段区间/cm^{-1}	识别率/%
320~540	44.67	1 035~1 185	63.13
590~685	58.00	1 240~1 400	68.47
820~925	54.13	1 415~1 515	68.47
925~990	73.00	1 595~1 710	88.80
1 000~1 035	67.47	1 730~1 790	42.27

2.2.3　小结

实验采集了不同品牌乳酪制品的拉曼光谱数据,与 k 近邻算法联用,研究讨论了不同谱图处理条件下鉴别算法的识别准确率。从上述分析结果可以看出,拉曼光谱数据未经处理时,鉴别算法识别率仅为 82.27%;经 coif1 小波降噪处理后,识别率可提升至 86%;经归一化处理至 [0,1] 区间后,识别率可达到 93.73%;再经 sym5 小波增强后,识别率可达到 94.4%;最后,经融合处理后,即选择 1 415~1 515 cm^{-1} 以及 1 595~1 710 cm^{-1} 拉曼光谱数据区间融合,识别率可达到 95.4%。由此可以看出,小波变换处理可以有效实现谱图数据的噪声滤除、特征信号增强。最终,本节建立起了一套基于谱图数据处理的鉴别算法优化流程,可为乳制品质量管理判别提供技术支持。

第3章 乳制品拉曼光谱数据特征提取研究

通过本书第2章的研究,我们可以获知,拉曼光谱数据的标准化、降噪等预处理对于揭示乳制品样品的质量判别规律有着重要的影响。本章将进一步通过实验案例研究光谱数据特征提取对乳制品质量判别的影响情况,主要聚焦于拉曼光谱的化学特征提取,通过拉曼光谱峰高、峰面积、峰比值、峰间关系等多方面比较探索,以期揭示光谱特性规律及其与不同判别方法间的关系。

3.1 基于拉曼光谱化学特征提取的乳制品质量判别研究

拉曼光谱可以提供测试样品诸多的分子振动信息,可用于目标分子的定性、定量和结构分析。在使用拉曼光谱法开展乳制品质量分析的过程中,常用以下两种方法。一种是参照朗伯比尔定律在一定范围内进行质量特性分子的浓度分析,如根据质量特性分子含量与拉曼光谱的峰强度或峰面积等之间的关系建立标准曲线。例如,Li 等研究建立了一种侧流结合表面增强拉曼光谱的奶粉中两种 β-内酰胺类抗生素的检定方法,通过使用峰高法和多元线性回归分析,可以实现头孢氨苄和氨苄青霉素的定量检测[1]。Zhang 等研究建立了一种基于表面增强拉曼光谱的乳制品中硫氰酸钠的测定方法。硫氰酸钠是一种保鲜剂,我国原卫生部于2008年发布的文件《食品中可能违法添加的非食用物质和易滥用的食品添加剂品种名单(第一批)》,明确规定乳及乳制品中禁止使用硫氰酸钠。通过建立特征峰标准曲线可实现硫氰酸钠的含量定量分析[2]。这种测定方法显示出特定峰及标值与目标分子间存在数理关系,但数据处理通常只是简单地使用特定峰值的信息,数据利用率相对较低。另一种是基于化学计量学算法的质量判别分析,如人

---

[1] Li X Z, Wang X N, Wang L Y, et al. Duplex detection of antibiotics in milk powder using lateral-flow assay based on surface-enhanced Raman spectroscopy[J]. Food Analytical Methods, 2021, 14(1): 165-171.

[2] Zhang Z Y, Liu J, Wang H Y. Microchip-based surface-enhanced Raman spectroscopy for the determination of sodium thiocyanate in milk[J]. Analytical Letters, 2015, 48(12): 1930-1940.

工神经网络和偏最小二乘法等[①②③]。主成分分析是一种常用的数学变换型特征提取法，通过正交变换将一组可能存在相关性的变量转换为一组线性不相关的变量，这些转换后的变量被称为主成分，使用少量的主成分即可代表原始数据的大量信息，可以减少冗余信息的干扰、达到降维的效果，可望提高判别算法的效率，但是主成分分析在其所提取特征的化学信息解析方面有一定的难度[④]。

为此，本节聚焦研究乳制品质量判别分析中的拉曼光谱化学特征提取法，其中主要包括以下三个方面。首先，传统的乳制品质量判别方法主要基于营养成分或非法成分含量的分析，在针对乳制品的品牌鉴别方面存在技术难点[⑤]，而本节所建立的方法在乳制品的品牌鉴别方面具有一定的应用潜力。其次，传统乳制品的鉴别分析常常基于相对复杂的特征提取，例如，有着数学变换特点的主成分分析法，而本节研究所探讨的是化学特征提取分析，比主成分分析更加简单和直接，有利于扩展拉曼光谱的潜在应用。最后，将光谱的特征提取与统计控制图方法进行了结合研究，采用较为直观的方式方法进行展示分析[⑥⑦]。

3.1.1 实验部分

1. 材料

实验用乳制品均购置于南京苏果超市，其中，鼎鑫奶片标记为品牌 P1，朴珍奶片标记为品牌 P2，雪原奶片标记为品牌 P3。各品牌乳制品的数量分别为 49 份、42 份和 60 份。

2. 仪器与设备

实验的拉曼光谱数据采集使用了便携式激光拉曼光谱仪，光谱仪型号：

---

① Alves da Rocha R, Paiva I M, Anjos V, et al. Quantification of whey in fluid milk using confocal Raman microscopy and artificial neural network[J]. Journal of Dairy Science, 2015, 98(6): 3559 - 3567.

② Mendes T O, Junqueira G M A, Porto B L S, et al. Vibrational spectroscopy for milk fat quantification: Line shape analysis of the Raman and infrared spectra[J]. Journal of Raman Spectroscopy, 2016, 47(6): 692 - 698.

③ Wang J P, Xie X F, Feng J S, et al. Rapid detection of Listeria monocytogenes in milk using confocal micro-Raman spectroscopy and chemometric analysis [J]. International Journal of Food Microbiology, 2015, 204: 66 - 74.

④ Cebi N, Dogan C E, Develioglu A, et al. Detection of l-Cysteine in wheat flour by Raman microspectroscopy combined chemometrics of HCA and PCA[J]. Food Chemistry, 2017, 228: 116 - 124.

⑤ Qi M H, Huang X Y, Zhou Y J, et al. Label-free surface-enhanced Raman scattering strategy for rapid detection of penicilloic acid in milk products[J]. Food Chemistry, 2016, 197(Pt A): 723 - 729.

⑥ Zhang Z Y, Gui D D, Sha M, et al. Raman chemical feature extraction for quality control of dairy products[J]. Journal of Dairy Science, 2019, 102(1): 68 - 76.

⑦ 王海燕,等.食药质量安全检测技术研究[M].北京:科学出版社,2023:1 - 215.

Prott-ezRaman-D3,厂家:美国恩威光电公司(Enwave Optronics)。随后,使用仪器自带的 SLSR Reader V8.3.9 软件对获取的样品拉曼光谱信号进行了基线校正。拉曼光谱仪的激光器激发波长为 785 nm,激光功率为 450 mW,照射时间为 50 s,光谱范围为 250~2 339 cm^{-1},光谱分辨率为 1 cm^{-1}。

3. 数据分析方法

(1) 欧氏距离

欧氏距离是一种常用的距离测量方法。计算 n 维空间中两个样本之间欧氏距离公式如(3.1)所示。

$$d = \sqrt{\sum_{i=1}^{n}(x_i - y_i)^2} \tag{3.1}$$

式中,d 表示欧氏距离,x_i 和 y_i 分别表示 x 和 y 样本的 i 维分量。在本节工作中,x_i 和 y_i 表示两个样本在一个特定波段的拉曼光谱强度(或峰面积,峰比值),i 表示拉曼光谱波数。

(2) 统计控制图

实验中,统计控制图选用的是单值移动极差控制图,其计算公式分述如下。对于单值(d)控制图,计算公式如(3.2)所示。

$$\begin{cases} \mathrm{UCL}_d = \bar{d} + 2.66\bar{R} \\ \mathrm{CL}_d = \bar{d} \\ \mathrm{LCL}_d = \bar{d} - 2.66\bar{R} \end{cases} \tag{3.2}$$

对于移动极差(moving range, MR)控制图,计算公式如(3.3)所示。

$$\begin{cases} \mathrm{UCL}_{\mathrm{MR}} = 3.267\bar{R} \\ \mathrm{CL}_{\mathrm{MR}} = \bar{R} \\ \mathrm{LCL}_{\mathrm{MR}} = 0 \end{cases} \tag{3.3}$$

式中,d 和 $\bar{d}$ 分别表示样本的欧氏距离和平均值;R 表示移动极差,即 $R = |d_{i+1} - d_i|$;d_i 表示样本 i 变量的欧氏距离;$\bar{R}$ 表示 R 的平均值;CL 即 center line,表示中心线;UCL 即 upper control limit,表示上控制限;LCL 即 lower control limit,表示下控制限。

(3) 拉曼光谱特征峰面积的计算。根据梯形法,特征峰面积计算如(3.4)所示。

$$\int_a^b f(x)\mathrm{d}x = \frac{b-a}{2N}\sum_{n=1}^{N}(f(x_n) + f(x_{n+1})) \tag{3.4}$$

式中,$f(x)$ 表示梯形函数,a 表示特征峰的初始波数,b 表示特征峰的尾端波数，N 表示初始与尾端之间的区间数（$N=b-a$），n 表示 a 和 b 段的光谱波数。使用的计算工具是 MATLAB R2013a。

3.1.2 结果与讨论

1. 乳制品的拉曼光谱分析

实验首先采集了奶片的拉曼光谱，如图 3-1 所示。根据已有的文献报道，主要谱峰可做如下归属[1][2][3]。在 1 752 cm^{-1} 处的拉曼谱峰可能主要源自脂肪酸的 C═O 伸缩振动，1 660 cm^{-1} 处的谱峰可能与蛋白质酰胺 CONH 的 C═O 伸缩振动和不饱和脂肪酸的 C═C 伸缩振动有关，在 1 462 cm^{-1} 处的谱峰可能是源自脂肪和碳水化合物分子的 CH$_2$ 变形振动，在 1 337 cm^{-1} 处的谱峰可能是碳水化合物分子的 C—O 伸缩振动和 C—O—H 变形振动，在 1 307 cm^{-1} 处的谱峰可归因于脂质分子的 CH$_2$ 扭曲振动，在 1 256 cm^{-1} 处的谱峰是由于碳水化合物分子的 CH$_2$ 扭曲振动。800 至 1 200 cm^{-1} 范围内约有 6 个主要谱峰，可认为主要是由碳水化合物引起的，如 C—C 伸缩振动和 C—O—H 变形振动（1 130 cm^{-1}、1 078 cm^{-1} 和 1027 cm^{-1}），C—O—C 变形振动（922 cm^{-1} 和 861 cm^{-1}）。此外，在 1 007 cm^{-1} 处的谱峰稍微特殊，可能主要源自蛋白质苯丙氨酸的苯环呼吸振动。250 至 800 cm^{-1} 范围内约有 7 个主要谱峰，分别是 C—C—O 变形振动（777 cm^{-1}）、C—S 伸缩振动（718 cm^{-1}）、C—C—C 变形振动和 C—O 扭曲振动（570 cm^{-1}）、葡萄糖（520 cm^{-1}）、C—C—C 变形振动和 C—O 扭曲振动（487 cm^{-1}）、葡萄糖（425 cm^{-1}）和乳糖（373 cm^{-1}）。图示不同品牌乳制品的拉曼光谱间较为相似，仅凭人眼难以区分。此外，从前述分析及结合已有研究工作可以看出，拉曼光谱可以表征出乳制品丰富的分子振动信息，且不同类型乳制品的拉曼光谱图示间有一定的相似性[4][5]。

---

[1] Almeida M R, de S Oliveira K, Stephani R, et al. Fourier-transform Raman analysis of milk powder: A potential method for rapid quality screening[J]. Journal of Raman Spectroscopy, 2011, 42(7): 1548-1552.

[2] Mazurek S, Szostak R, Czaja T, et al. Analysis of milk by FT-Raman spectroscopy[J]. Talanta, 2015, 138: 285-289.

[3] McGoverin C M, Clark A S S, Holroyd S E, et al. Raman spectroscopic quantification of milk powder constituents[J]. Analytica Chimica Acta, 2010, 673(1): 26-32.

[4] Nunes P P, Almeida M R, Pacheco F G, et al. Detection of carbon nanotubes in bovine raw milk through Fourier transform Raman spectroscopy[J]. Journal of Dairy Science, 2024, 107(5): 2681-2689.

[5] Batesttin C, Ângelo F F, Rocha R A, et al. High resolution Raman spectroscopy of raw and UHT bovine and Goat milk[J]. Measurement: Food, 2022, 6: 100029.

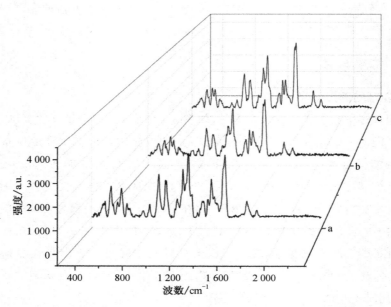

图 3-1　不同品牌 P1(a)、P2(b)和 P3(c)乳制品的拉曼光谱图

2. 基于拉曼峰值强度的乳制品特征提取与分析

拉曼光谱的出峰位置与特定的乳制品组分分子振动间有着密切的关系,与此同时,峰的形状也蕴含着深刻内涵,故首先从峰强度这一参数开展进一步的研究。以 P1 品牌乳制品为主要研究对象,考察基于拉曼峰强为特征的质量判别可行性。从产品质量波动的一般规律而言,稳定的产品其质量应该是在一个合理的统计区间范围内正常波动,异常产品即其他品牌乳制品则不在此可控范围内,据此可望实现样品间的区分判别。研究以收集到的样品拉曼光谱全谱数据(250～2 339 cm^{-1})作为输入,为了定量化地进行样品描述,本节使用了欧氏距离法。欧氏距离法可以评估两个样本之间的相似性:当样本之间的相似性较高时,欧氏距离数值较小;相反,当样本之间的相似性较低时,欧氏距离数值较大[1]。将收集到的 P1 品牌乳制品的所有拉曼光谱的平均谱作为样品理论真值的最佳估计,然后计算每个实验样品与平均值之间的欧氏距离。随后,使用计算得到的欧氏距离构建 P1 品牌乳制品的单值和移动极差控制图,用以评估样品的质量稳定性,如图

---

[1] Chen J B, Zhou Q, Noda I, et al. Quantitative classification of two-dimensional correlation spectra[J]. Applied Spectroscopy, 2009, 63(8): 920-925.

3-2所示。计算出的各个欧氏距离数值在中心线周围波动,表明样本之间存在一定的质量波动。根据正态分布模型,计算控制图的上、下限,结果表明,虽然样品的欧氏距离值一直在波动变化,但它们都处在控制限的范围内,反映出当采用全光谱作为特征输入时,P1品牌乳制品各样品间具有较高的稳定一致性。进一步,计算P2品牌乳制品、P3品牌乳制品(对照组)和P1品牌乳制品(实验组)均值之间的欧氏距离,如图3-3所示。可以看出,5个P2品牌乳制品样品低于控制限(UCL=5 345),2个P3品牌乳制品样品接近控制限(UCL=5 345)。由图3-2可以看出以全光谱作为特征输入,同一品牌的P1乳制品存在质量稳定性,但由图3-3结果显示出不同品牌的乳制品间尚不能有效区分,原因可能是由于不同品牌样品的拉曼光谱较为相似,仅凭全光谱强度区分度不够,以及光谱可能存在噪声和冗余信息。故有必要进一步开展样品拉曼光谱的化学特征提取研究,探讨提高光谱数据的判别性能。

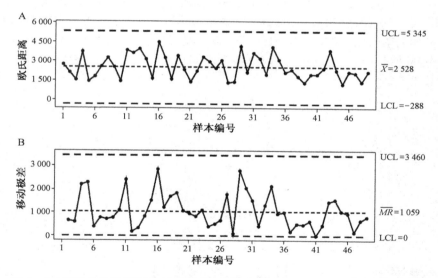

图3-2 基于拉曼峰值强度(全光谱)欧氏距离的P1乳制品的质量波动单值(A)和移动极差(B)控制图

注:UCL表示上控制限,LCL表示下控制限,$\overline{X}$表示单值的均值,$\overline{MR}$表示移动极差的均值。

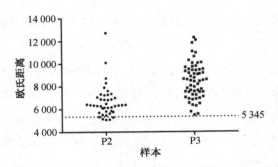

图3-3 基于拉曼峰值强度(全光谱)的 P2 乳制品、P3 乳制品和 P1 乳制品(均值)之间的欧氏距离运算结果情况

通过对样品拉曼光谱的仔细比较分析,进一步选取 1 307 cm^{-1} 处的拉曼强度,以 P1 品牌乳制品为主要研究对象,开展进一步的欧氏距离计算和统计控制图绘制分析,结果如图 3-4 所示。可以看到,与图 3-2 相比,乳制品 P1 样品之间的欧氏距离显著减小,并且围绕中心线波动,没有样本跃出控制限,这一结果显示出,同一品牌乳制品各样品之间具有较高的相似性。采用相同的操作方法计算 P2、P3 和 P1 乳制品(均值)之间的欧氏距离,如图 3-5 所示。可以看出,12 份 P2 乳制品样品落入控制限(UCL=169.1),而 P3 乳制品则均远离控制限,显示出良好的分离效果。从以上研究可以看出,仅以拉曼光谱强度作为特征输入时,质量判别效果不能完全实现。

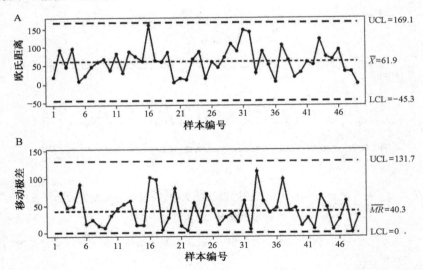

图3-4 基于 1 307 cm^{-1} 处拉曼峰强度欧氏距离的 P1 乳制品的质量波动单值(A)和移动极差(B)控制图

注:UCL 表示上控制限,LCL 表示下控制限,$\overline{X}$ 表示单值的均值,$\overline{MR}$ 表示移动极差的均值。

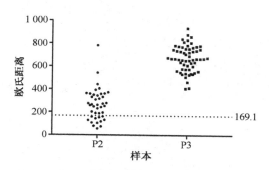

图 3-5 基于 1 307 cm^{-1} 处拉曼峰强度的乳制品 P2、乳制品 P3 和乳制品 P1(均值)之间的欧氏距离运算结果情况

3. 基于拉曼峰面积的乳制品特征提取与分析

峰面积也是描述拉曼光谱峰的一个重要特征参数,可以反映出样品中丰富的特征分子信息,峰面积涉及一个光谱距离,所以它通常包含在峰值范围内的多个特征分子相互关联的信息或称含量信息。首先使用峰面积(全光谱,250~2 339 cm^{-1})作为特征输入,计算 P1 乳制品各样品与其均值之间的欧氏距离,绘制出单值移动极差控制图,如图 3-6 所示。结果显示出,所有样品均围绕中心线波动,并都在控制范围内。在相同的计算条件下,同步得到了 P2、P3 和 P1 乳制品均值之间的欧氏距离,如图 3-7 所示。结果显示出,37 个 P2 乳制品样品和 45 个 P3 乳制品样品进入控制限(UCL=136 543)内,表明依靠全谱峰面积运算无法区分不同品牌的样品。经过光谱分析,进一步选择 1 288~1 319 cm^{-1} 范围内的谱峰面积作为特征输入,按照前述方法,计算出 P1 乳制品各样品与其均值之间的欧氏距离,绘图如图 3-8 所示,结果显示出有 1 个样本跃出了控制限,这说明可能存在质量风险,不过其余的样品都在控制限内,总体来看,P1 乳制品的整体质量水平仍可认为处于可控状态。图 3-9 为相同条件下 P2、P3 和 P1 乳制品均值间的判别分析结果,可以看出,所有乳制品(对照组)均跃出了控制限,只有少量 P2 乳制品样本较为接近控制限(UCL=3 386)。结果显示出,谱峰面积作为一个特征可以表征出样本更丰富的差异信息,不过在以全谱面积作为特征输入时不能满足质量鉴别应用的需要,通过选取适当的特征峰面积为输入,有望提高判别模型的区分效果。

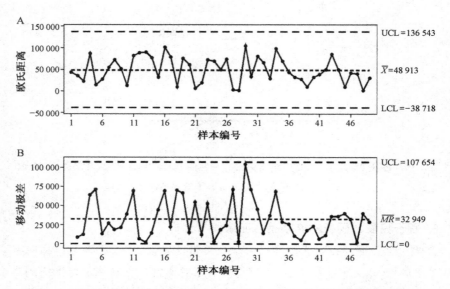

图 3-6 基于拉曼峰面积(全光谱)欧氏距离的 P1 乳制品的质量波动单值(A)和移动极差(B)控制图

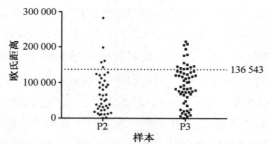

图 3-7 基于拉曼峰面积(全光谱)的乳制品 P2、乳制品 P3 和乳制品 P1(均值)之间的欧氏距离运算结果情况

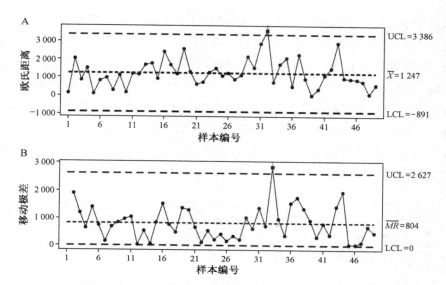

图3-8 基于 1 288～1 319 cm^{-1} 区间拉曼峰面积欧氏距离的 P1 乳制品质量波动单值(A)和移动极差(B)控制图

注：UCL 表示上控制限，LCL 表示下控制限，$\overline{X}$ 表示单值的均值，$\overline{MR}$ 表示移动极差的均值。

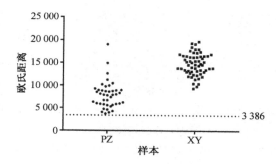

图3-9 基于 1 288～1 319 cm^{-1} 区间拉曼峰面积的 P2 乳制品、P3 乳制品和 P1 乳制品(均值)之间的欧氏距离运算结果情况

4. 基于拉曼峰比值的乳制品特征提取与分析

峰比值也是一个描述拉曼光谱峰的重要特征参数，可以表征谱峰间的联动变化关系[①]。在同品牌乳制品的生产过程中，其原料配比和生产工艺一般是保持稳定的，故此类产品的峰比值常保持在一定波动范围。选取 1 337 cm^{-1}/1 307 cm^{-1}

---

① 王霁月，陆乃彦，季莹，等. 人乳与市售婴儿配方乳粉脂质比较研究[J]. 食品安全质量检测学报，2020，11(21)：7784-7790.

处光谱峰强度比值作为特征输入,计算每个 P1 品牌乳制品样品与其平均值之间的欧氏距离,运算结果绘制控制图得到图 3-10。可见,P1 乳制品各样品与均值间的欧氏距离围绕着中心线附近波动,都落在控制限范围内,显示出样品具有较高的稳定性和一致性。图 3-11 为样品在 P2、P3 和 P1 乳制品(均值)之间的关系,显示出 P2 和 P3 乳制品的欧氏距离远离了 P1 乳制品的控制限(UCL=0.394 5),实现了乳制品对照组与实验组的有效区分。不同品牌的乳制品间配方存在差异,表现在原料含量及比值的差异,通过谱峰之间的比值分析,显示出较高的区分效果。

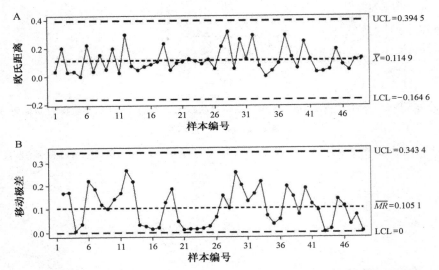

图 3-10 基于拉曼峰比值($1\,337\,cm^{-1}/1\,307\,cm^{-1}$)的 P1 乳制品质量波动单值(A)和移动极差控制图(B)

注:UCL 表示上控制限,LCL 表示下控制限,$\overline{X}$ 表示单值的均值,$\overline{MR}$ 表示移动极差的均值。

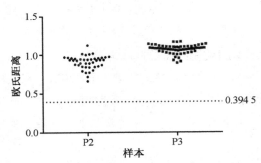

图 3-11 根据 $1\,337\,cm^{-1}/1\,307\,cm^{-1}$ 拉曼峰比值的 P2 乳制品、P3 乳制品和 P1 乳制品(均值)之间的欧氏距离运算结果情况

5. 基于拉曼光谱多重特征的乳制品质量鉴别分析

通过上述分析,本节提出了一种基于拉曼峰强度、峰面积和峰比值的特征提取方法,并进一步从三维空间进行了实验样品的鉴别分析。研究依据峰强度(1 307 cm^{-1})、峰面积(1 288~1 319 cm^{-1})、峰比值(1 337 cm^{-1}/1 307 cm^{-1})特征分别计算了各品牌乳制品样品与P1乳制品(均值)之间的欧氏距离,计算结果分别作为 x 轴、y 轴和 z 轴,绘制出样品的三维空间分布图,如图 3-12 所示。可以看出,实验组 P1 乳制品与对照组 P2、P3 乳制品在三维特征空间中存在明显差异,可以被有效地分离出来。因此,研究成功地建立了一种基于拉曼光谱化学特征提取的乳制品判别方案,并可为相关食品质量鉴别提供技术参考。

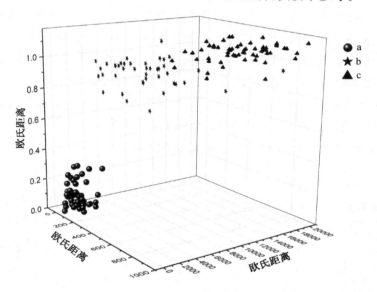

图 3-12 不同品牌 P1(a)、P2(b)和 P3(c)乳制品基于拉曼光谱的峰强度(x 轴)、峰面积(y 轴)和峰比值(z 轴)的欧氏距离三维图

3.1.3 小结

拉曼光谱技术能够快速地表征出实验样品丰富的分子振动信息,是乳制品快速分析和质量识别的有力工具。光谱特征提取可以提高判别方法的效率,与传统的数学变换算法相比,化学特征提取方法有利于化学信息的分析,更为直接,可加深人们对提取特征的认知。研究以拉曼峰强度、峰面积和峰比值作为化学特征输入,结合欧氏距离,并结合统计控制图分析,结果呈现如下:(1) 使用特征提取的方案,其判别效果优于直接使用全谱作为特征输入;(2) 拉曼光谱峰强度、谱面积

和谱比值在作为化学特征输入时,它们的判别效能也存在差异,在本实验中表现出依次增加的情况;(3)研究提取的化学特征可以用于乳制品的质量判别,研究方法可供相关食品化学特征提取所参考。

3.2 基于拉曼光谱特征提取与融合的乳制品统计判别

乳制品作为食品的重要组成部分,其质量安全一直是普通消费者、监管机构和科学家关注的热点问题之一。乳制品的安全性评价主要涉及非法添加剂(如三聚氰胺、硫氰酸钠)①、污染物(如铅、汞、砷、铬、硝酸盐)②、真菌毒素(如黄曲霉毒素 B1、黄曲霉毒素 M1)③④、微生物(如金黄色葡萄球菌、沙门氏菌)⑤、药物残留(环丙沙星)⑥等。乳制品的质量评价主要涉及产品的波动性和一致性,即产品应保持较高稳定性⑦。质量的定义是与变化成反比⑧。在以往的文献报道中,研究主要集中在乳制品的安全因素上,如研究微量有害物质的定性和定量测定⑨。然而,现有的研究方法也面临着一些挑战,尤其是在质量评价领域⑩。例如,近年来,我国执法人员查获了一些制造和销售假冒食品的案件,而新的情况是,这些假冒产品是一些低质低价的普通产品,符合国家相关质量安全标准要求,对人体并

---

① Kaleem A, Azmat M, Sharma A, et al. Melamine detection in liquid milk based on selective porous polymer monolith mediated with gold nanospheres by using surface enhanced Raman scattering[J]. Food Chemistry, 2019, 277: 624-631.

② Singh P, Singh M K, Beg Y R, et al. A review on spectroscopic methods for determination of nitrite and nitrate in environmental samples[J]. Talanta, 2019, 191: 364-381.

③ Xue Z H, Zhang Y X, Yu W C, et al. Recent advances in aflatoxin B1 detection based on nanotechnology and nanomaterials-a review[J]. Analytica Chimica Acta, 2019, 1069: 1-27.

④ Qu L L, Jia Q, Liu C Y, et al. Thin layer chromatography combined with surface-enhanced Raman spectroscopy for rapid sensing aflatoxins[J]. Journal of Chromatography A, 2018, 1579: 115-120.

⑤ Stöckel S, Kirchhoff J, Neugebauer U, et al. The application of Raman spectroscopy for the detection and identification of microorganisms[J]. Journal of Raman Spectroscopy, 2016, 47(1): 89-109.

⑥ 韩斯琴高娃, 孙佳, 包琳, 等. 基于 SERS 技术检测牛奶中环丙沙星的研究[J]. 药物分析杂志, 2018, 38(5): 802-805.

⑦ Montgomery D C. Introduction to Statistical Quality Control[M]. 6th ed. Hoboken: John Wiley & Sons, Inc., 2009: 1-734.

⑧ Montgomery D C. Introduction to statistical quality control[M]. 7th ed. Hoboken: John Wiley & Sons, Inc., 2013.

⑨ Hu X T, Shi J Y, Shi Y Q, et al. Use of a smartphone for visual detection of melamine in milk based on Au@Carbon quantum dots nanocomposites[J]. Food Chemistry, 2019, 272: 58-65.

⑩ Wang X, Esquerre C, Downey G, et al. Assessment of infant formula quality and composition using Vis-NIR, MIR and Raman process analytical technologies[J]. Talanta, 2018, 183: 320-328.

无伤害,属于合格产品,但造假者却可以利用它们冒充高质产品,利用相关差价达到获取非法利益的目的。对于乳制品而言,不同品牌的乳制品具有相似的外观和成分,而这成为高效识别的挑战之一[1][2][3]。

以不同品牌的巴氏杀菌乳为例,从外观上来看,它们都是乳白色的产品,人眼直接识别是不可行的,这提示我们需要开展更趋智能的乳制品质量鉴定研究工作。质量判别的研究可分为两个方面:质量表征和统计判别。可用于乳制品质量表征的技术包括色谱法、质谱法和光谱学等。色谱和质谱检测通常需要对样品进行预处理,这既费时又费力[4]。光谱检测方法包括紫外可见光谱法、荧光光谱法、红外光谱法和拉曼光谱法等。紫外可见光谱和荧光光谱可以提供相对较少的关于乳制品的信息[5],红外光谱对水分子较为敏感,与前述光谱相比,拉曼光谱具有许多优点,包括:(1)可以直接检测酸奶样品;(2)水分子拉曼散射截面小,不影响样品检测;(3)便携式拉曼光谱仪可实现现场样品检测;(4)拉曼光谱可以表征样品的丰富分子振动信息并且可以提供样品的指纹特性[6]。因此,基于拉曼光谱的乳制品质量鉴别研究成为研究热点之一。每个样本的拉曼光谱是一个数字矩阵,它可以被用作样本质量特性,以输入到随后的统计模型中。

统计判别是高效、深入挖掘表征数据内在规律,实现乳制品质量科学控制的重要技术。在特征提取方面,现有的方法主要包括主成分分析等。由于主成分分析是一种数学变换特征提取方法,它可以实现数据降维,不过在样品拉曼光谱的化学信息解析方面尚存在挑战[7]。因此,本节在收集和分析乳制品拉曼光谱的基础上,重点聚焦分析拉曼光谱特征数据的化学特征提取和融合,结合支持向量机

---

[1] Zhang Y S, Zhang Z Y, Zhao Y J, et al. Adaptive compressed sensing of Raman spectroscopic profiling data for discriminative tasks[J]. Talanta, 2020, 211: 120681.

[2] Karacaglar N N Y, Bulat T, Boyaci I H, et al. Raman spectroscopy coupled with chemometric methods for the discrimination of foreign fats and oils in cream and yogurt[J]. Journal of Food and Drug Analysis, 2019, 27(1): 101-110.

[3] McGoverin C M, Clark A S S, Holroyd S E, et al. Raman spectroscopic quantification of milk powder constituents[J]. Analytica Chimica Acta, 2010, 673(1): 26-32.

[4] Slimani K, Pirotais Y, Maris P, et al. Liquid chromatography-tandem mass spectrometry method for the analysis of N-(3-aminopropyl)-N-dodecylpropane-1, 3-diamine, a biocidal disinfectant, in dairy products[J]. Food Chemistry, 2018, 262: 168-177.

[5] Xiong S Q, Adhikari B, Chen X D, et al. Determination of ultra-low milk fat content using dual-wavelength ultraviolet spectroscopy[J]. Journal of Dairy Science, 2016, 99(12): 9652-9658.

[6] Tan Z, Lou T T, Huang Z X, et al. Single-drop Raman imaging exposes the trace contaminants in milk[J]. Journal of Agricultural and Food Chemistry, 2017, 65(30): 6274-6281.

[7] Weng S Z, Yuan H C, Zhang X Y, et al. Deep learning networks for the recognition and quantitation of surface-enhanced Raman spectroscopy[J]. The Analyst, 2020, 145(14): 4827-4835.

算法进行统计结果分析,综合提出了一种新的判别分析思路[①]。

3.2.1 实验部分

1. 样品和仪器

实验用乳制品均购置于南京苏果超市,其中,莫斯利安巴氏热处理风味酸奶产品标记为品牌 P1,安慕希巴氏热加工风味酸奶产品标记为品牌 P2,纯甄巴氏热处理风味酸奶产品标记为品牌 P3。这些样品的外观呈乳白色,每个品牌有 30 个样品。

样品的拉曼光谱信号采集使用便携式激光拉曼光谱仪,光谱仪型号:Prott-ezRaman-D3,厂家:美国恩威光电公司(Enwave Optronics)。激光器的激发波长为 785 nm,激光功率 450 mW,激光光斑直径约为 100 μm,照射时间为 150 s。光谱范围为 255~1 974 cm^{-1},分辨率为 1 cm^{-1}。样品在实验过程中没有进行额外的物理和化学处理。

2. 数据分析

对收集到的实验样品拉曼光谱信号进行基线校正,使用的是光谱仪自带的软件 SLSR Reader V8.3.9 来完成。使用 MATLAB 软件(美国 MathWorks 公司)执行用于拉曼光谱表征数据的特征提取、融合判别以及支持向量机算法运算。

对于支持向量机算法,基本思想如下所示[②③④]:

对于已知的训练集(T)

$$T = \{(x_1, y_1), \cdots, (x_m, y_m)\} \in (X \times Y)^m$$

在该公式中,$x_i \in X = \mathbf{R}^n, y_i \in Y = \{1, -1\}(i = 1, 2, \cdots, m)$;$x_i$ 是拉曼光谱数据,y_i 是类别标签。这是一个两类模型,对于多分类问题,可以通过构造多个二分类支持向量机来解决。

通过选择适当的核函数 $K(x, x')$ 和参数 C,构建并求解优化问题:

[①] Zhang Z Y. The statistical fusion identification of dairy products based on extracted Raman spectroscopy[J]. RSC Advances, 2020, 10(50): 29682 - 29687.

[②] Ren S X, Gao L. Improvement of the prediction ability of multivariate calibration by a method based on the combination of data fusion and least squares support vector machines[J]. Analyst, 2011, 136(6): 1252 - 1261.

[③] Sun H T, Lv G D, Mo J Q, et al. Application of KPCA combined with SVM in Raman spectral discrimination[J]. Optik, 2019, 184: 214 - 219.

[④] Bakhtiaridoost S, Habibiyan H, Muhammadnejad S, et al. Raman spectroscopy-based label-free cell identification using wavelet transform and support vector machine[J]. RSC Advances, 2016, 6(55): 50027 - 50033.

$$\min_{\alpha} \frac{1}{2}\sum_{i=1}^{j}\sum_{j=1}^{m} y_i y_j \alpha_i \alpha_j K(x_i,x_j) - \sum_{j=1}^{m}\alpha_j$$

$$s.t. \sum_{i=1}^{m} y_i\alpha_i = 0, \qquad 0 \leqslant \alpha \leqslant C, \qquad i=1,\cdots,m$$

获得最佳解决方案：

$$\alpha^* = (\alpha_1^*,\cdots,\alpha_m^*)^{\mathrm{T}}$$

选择了 α^* 的正分量 $0 < \alpha_j^* < C$，并且相应地计算阈值：

$$b^* = y_j - \sum_{i=1}^{m} y_i\alpha_i^* K(x_i - x_j)$$

构建决策函数：

$$f(x) = \mathrm{sgn}\Big(\sum_{i=1}^{m}\alpha_i^* y_i K(x \cdot x_i) + b^*\Big)$$

本节采用径向基函数作为核函数：

$$K(x,x_i) = \exp\Big[-\frac{\parallel x - x_i \parallel^2}{2\sigma^2}\Big]$$

其中，σ 表示径向基函数的核宽度。

3.2.2 结果和讨论

1. 乳制品的拉曼光谱表征分析

如图 3-13 所示，收集了三种乳制品的拉曼光谱，参考已有的相关文献报道，可以对它们的主要拉曼峰进行归属分析[1][2]。例如，1 759 cm^{-1} 处的谱峰可以归属于脂肪酸的 C=O 伸缩振动，1 468 cm^{-1} 处的谱峰可归因于脂肪和糖的 CH$_2$ 变形振动，1 016 cm^{-1} 处的谱峰比较特殊，可归属于苯环呼吸振动，其主要来源于蛋白质的苯丙氨酸。关于拉曼峰归属的更多信息如表 3-1 所示。由图 3-13 可以看出，每个实验样品的拉曼光谱主要出峰数量接近 20 个，并且每个峰的宽度都很窄，具有较强的指纹特征。然而，我们也可以看到，各品牌实验样品间的拉曼光谱在峰位置、峰形等方面非常相似。

基于拉曼峰 1 468 cm^{-1} 处的强度值，运用单值移动极差控制图进行实验样品的质量波动评估分析，结果显示出，每个品牌乳制品的光谱强度值变化都是处于

---

[1] Zhang Z Y, Gui D D, Sha M, et al. Raman chemical feature extraction for quality control of dairy products[J]. Journal of Dairy Science, 2019, 102(1): 68-76.
[2] 王筠钠,李妍,李扬,等.光谱学技术在稀奶油乳脂肪研究中的应用[J].光谱学与光谱分析,2019, 39(6):1773-1778.

统计可控的质量波动范围内,P1、P2 和 P3 乳制品的拉曼光谱强度分别位于 462.6~757.8,397.4~682.2 和 482.5~739.8 的范围内。实验结果表明,光谱数据符合一定的统计规律,同时,也揭示出各品牌样品间光谱强度存在交集,仅靠光谱强度分析无法实现品牌识别。此外,三种乳制品的营养成分也较为接近,如表 3-2 所示,传统的成分鉴定方法难以奏效,也存在成分分析较为费时费力的挑战。为此,以它们的判别分析为例,进一步开展一种新的统计判别方法研究。

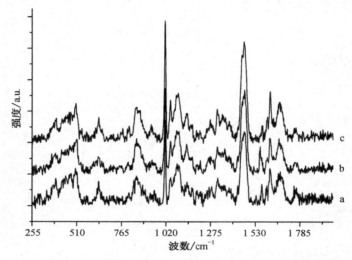

图 3-13　不同品牌 P1(a)、P2(b)和 P3(c)乳制品的拉曼光谱

表 3-1　巴氏杀菌乳主要拉曼峰及其各自的可能归属

波数/cm^{-1}	归属
1 759	$\nu(C{=}O)_{酯}$
1 670	$\nu(C{=}O)$(酰胺Ⅰ),$\nu(C{=}C)$
1 615	$\nu(C{-}C)_{环}$
1 566	$\delta(N{-}H)$;$\nu(C{-}N)$(酰胺Ⅱ)
1 468	$\delta(CH_2)$
1 315	$\tau(CH_2)$
1 279	$\gamma(CH_2)$
1 139	$\nu(C{-}O)+\nu(C{-}C)+\delta(C{-}O{-}H)$
1 085	$\nu(C{-}O)+\nu(C{-}C)+\delta(C{-}O{-}H)$
1 046	$\nu(C{-}O)+\nu(C{-}C)+\delta(C{-}O{-}H)$

续表 3-1

波数/cm^{-1}	归属
1 016	苯环呼吸振动(苯丙氨酸),ν(C—C)$_{环}$
953	δ(C—O—C)+δ(C—O—H)+ν(C—O)
851	δ(C—C—H)+δ(C—O—C)
807	δ(C—C—O)
633	δ(C—C—O)
502	葡萄糖
445	δ(C—C—H)+τ(C—O)
389	乳糖

注：ν—伸缩振动，δ—变形振动，τ—弯曲振动，γ—面外弯曲振动。

表 3-2　不同品牌乳制品的主要成分含量

组分	品牌 P1	品牌 P2	品牌 P3
蛋白质	2.8 g/100 g	3.1 g/100 g	2.8 g/100 g
脂肪	3.0 g/100 g	3.1 g/100 g	3.2 g/100 g
糖类	12.5 g/100 g	13.0 g/100 g	12.5 g/100 g

2. 乳制品拉曼光谱的数据预处理和特征提取

首先，使用原始拉曼全光谱数据进行乳制品分类判别，基于支持向量机算法的平均识别率为 58%。测试条件是随机选择 2/3 的样本作为训练集，其余 1/3 的样本作为验证集，进行 10 次随机测试以获得平均识别准确率。特征提取分析的目的是减少冗余信息对判别模型的干扰，改善样本之间的细节差异，从而实现不同样本的准确判别，并有望节省计算运行时间。本节分别采用移动窗口法和谱带人工选择法对拉曼光谱区间进行筛选。移动窗口的宽度设置为 20 个波数，谱峰人工筛选的依据是图 3-13 各样品拉曼出峰情况人工划分光谱区间。实验结果分别如表 3-3 和 3-4 所示。所有运算结果均小于 70%，其中，基于波段 1 755~1 774 cm^{-1} 的平均识别率最高可达 69%，高于基于全谱的识别率，显示出特征提取有望提高判别效果。

表 3-3 基于拉曼光谱和支持向量机判别算法的乳制品判别结果（移动窗口间隔 $20\ \text{cm}^{-1}$）

光谱范围/cm^{-1}	识别率/%	光谱范围/cm^{-1}	识别率/%
255～274	42	1 115～1 134	48
275～294	57	1 135～1 154	49
295～314	46	1 155～1 174	44
315～334	47	1 175～1 194	55
335～354	56	1 195～1 214	54
355～374	39	1 215～1 234	46
375～394	36	1 235～1 254	54
395～414	43	1 255～1 274	44
415～434	48	1 275～1 294	46
435～454	54	1 295～1 314	64
455～474	46	1 315～1 334	53
475～494	44	1 335～1 354	41
495～514	52	1 355～1 374	43
515～534	55	1 375～1 394	43
535～554	44	1 395～1 414	37
555～574	35	1 415～1 434	49
575～594	42	1 435～1 454	47
595～614	50	1 455～1 474	51
615～634	55	1 475～1 494	49
635～654	62	1 495～1 514	58
655～674	42	1 515～1 534	58
675～694	54	1 535～1 554	50
695～714	56	1 555～1 574	52
715～734	42	1 575～1 594	66
735～754	50	1 595～1 614	53
755～774	57	1 615～1 634	58
775～794	51	1 635～1 654	56
795～814	47	1 655～1 674	49

续表 3-3

光谱范围/cm^{-1}	识别率/%	光谱范围/cm^{-1}	识别率/%
815～834	58	1 675～1 694	49
835～854	47	1 695～1 714	56
855～874	43	1 715～1 734	47
875～894	45	1 735～1 754	53
895～914	52	**1 755～1 774**	**69**
915～934	50	1 775～1 794	43
935～954	47	1 795～1 814	60
955～974	49	1 815～1 834	54
975～994	50	1 835～1 854	63
995～1 014	47	1 855～1 874	60
1 015～1 034	62	1 875～1 894	60
1 035～1 054	57	1 895～1 914	60
1 055～1 074	48	1 915～1 934	56
1 075～1 094	37	1 935～1 954	57
1 095～1 114	51	1 955～1 974	55

表 3-4 基于乳制品拉曼光谱和支持向量机判别算法的乳制品判别结果(拉曼波段人工筛选)

光谱范围/cm^{-1}	识别率/%	光谱范围/cm^{-1}	识别率/%
350～405	35	1 155～1 185	51
485～540	61	1 185～1 230	50
590～670	47	1 230～1 300	52
780～820	52	1 300～1 415	59
820～915	57	1 415～1 520	55
915～985	44	1 550～1 580	55
985～1 030	56	1 580～1 605	62
1 030～1 060	53	1 605～1 640	59
1 060～1 115	41	1 640～1 730	49
1 115～1 155	41	1 730～1 800	57

接下来,研究探讨拉曼光谱数据预处理方法对乳制品判别的影响。一般而言,光谱预处理方法有望帮助我们突出数据差异,提高分类算法的识别率。如图3-14至3-18所示,分别对拉曼光谱数据进行了一阶导数、二阶导数、三阶导数、四阶导数和归一化处理。通过直观的视觉判别分析,可以看出这些光谱之间的差异不是很明显,然而,支持向量机算法运算得到的平均识别率有一些变化,结果如表3-5所示。基于导数处理的判别效果有所下降,而基于归一化处理的判别效果显著增加。结果显示出,归一化操作对支持向量机算法有很大的影响,并有效地消除了原始光谱数据的量纲影响[1][2][3]。此外,随机选择2/3的样本数据作为训练集,其余1/3的数据作为验证集,在10次随机实验中研究移动窗口选择和谱带特征人工选择的平均判别效果,结果分别如表3-6和表3-7所示。移动窗口法在$335 \sim 354 \text{ cm}^{-1}$、$435 \sim 454 \text{ cm}^{-1}$和$835 \sim 854 \text{ cm}^{-1}$拉曼光谱范围内的识别准确率均超过70%。基于$485 \sim 540 \text{ cm}^{-1}$、$1\,155 \sim 1\,185 \text{ cm}^{-1}$、$1\,300 \sim 1\,415 \text{ cm}^{-1}$和$1\,415 \sim 1\,520 \text{ cm}^{-1}$拉曼光谱范围的平均识别率也分别不低于70%,基于$820 \sim 915 \text{ cm}^{-1}$区间的识别率达到83%。结果表明,与上述相应的基于原始数据(表3-3和3-4)的运算结果相比,归一化后基于特征谱带的识别率大部分得到了提高,不同谱带对支持向量机算法准确判别的贡献有所不同,不过识别率尚低于全谱数据(91%,表3-5),这可能是由于支持向量机算法缺乏足够的谱数据输入。如图3-19所示,样本之间的光谱特征区间存在一定差异。

---

[1] Tian F M, Tan F, Li H. An rapid nondestructive testing method for distinguishing rice producing areas based on Raman spectroscopy and support vector machine[J]. Vibrational Spectroscopy, 2020, 107: 103017.

[2] Chen H Z, Xu L L, Ai W, et al. Kernel functions embedded in support vector machine learning models for rapid water pollution assessment via near-infrared spectroscopy[J]. The Science of the Total Environment, 2020, 714: 136765.

[3] Li W L, Yan X, Pan J C, et al. Rapid analysis of the Tanreqing injection by near-infrared spectroscopy combined with least squares support vector machine and Gaussian process modeling techniques[J]. Spectrochimica Acta Part A, Molecular and Biomolecular Spectroscopy, 2019, 218: 271-280.

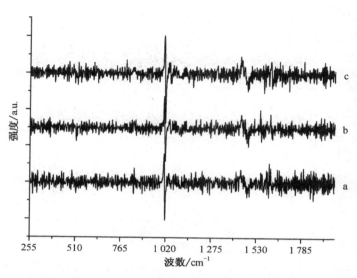

图 3-14 不同品牌 P1(a)、P2(b) 和 P3(c) 乳制品一阶导数处理后的拉曼光谱

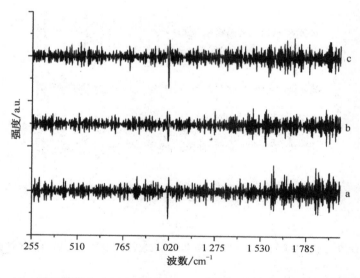

图 3-15 不同品牌 P1(a)、P2(b) 和 P3(c) 乳制品二阶导数处理后的拉曼光谱

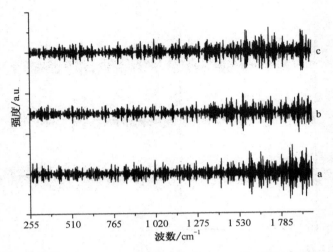

图 3-16 不同品牌 P1(a)、P2(b) 和 P3(c) 乳制品三阶导数处理后的拉曼光谱

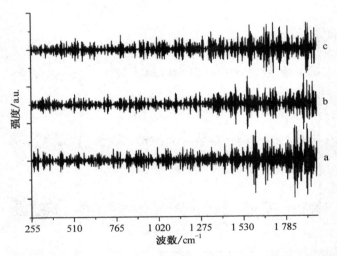

图 3-17 不同品牌 P1(a)、P2(b) 和 P3(c) 乳制品四阶导数处理后的拉曼光谱

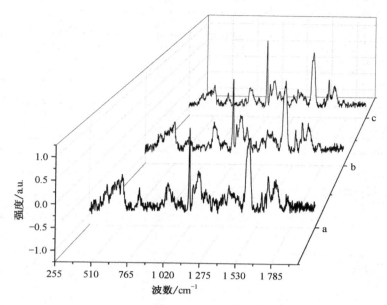

图 3-18 不同品牌 P1(a)、P2(b)和 P3(c)乳制品归一化处理后的拉曼光谱

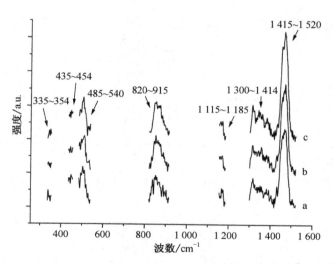

图 3-19 不同品牌 P1(a)、P2(b)和 P3(c)乳制品拉曼光谱特征提取区间

表 3-5 基于不同预处理方法和支持向量机判别算法的乳制品判别结果

光谱预处理方法	平均准确率/%	运行时间/s
原始数据	58	820
一阶导数	53	833
二阶导数	51	841
三阶导数	40	831
四阶导数	53	838
归一化	91	823

拉曼光谱的范围:255~1 974 cm^{-1}。

表 3-6 基于归一化后的拉曼光谱和支持向量机判别算法的乳制品判别结果
（移动窗口间隔 20 cm^{-1}）

光谱范围/cm^{-1}	识别率/%	光谱范围/cm^{-1}	识别率/%
255~274	44	1 115~1 134	61
275~294	57	1 135~1 154	60
295~314	53	1 155~1 174	57
315~334	48	1 175~1 194	55
335~354	**71**	1 195~1 214	53
355~374	51	1 215~1 234	41
375~394	53	1 235~1 254	57
395~414	57	1 255~1 274	50
415~434	55	1 275~1 294	60
435~454	**71**	1 295~1 314	55
455~474	61	1 315~1 334	60
475~494	65	1 335~1 354	53
495~514	57	1 355~1 374	51
515~534	65	1 375~1 394	51
535~554	48	1 395~1 414	53
555~574	46	1 415~1 434	62
575~594	46	1 435~1 454	62

续表 3-6

光谱范围/cm^{-1}	识别率/%	光谱范围/cm^{-1}	识别率/%
595~614	52	1 455~1 474	59
615~634	61	1 475~1 494	57
635~654	64	1 495~1 514	57
655~674	42	1 515~1 534	53
675~694	40	1 535~1 554	53
695~714	55	1 555~1 574	69
715~734	45	1 575~1 594	58
735~754	51	1 595~1 614	58
755~774	50	1 615~1 634	64
775~794	55	1 635~1 654	60
795~814	43	1 655~1 674	58
815~834	59	1 675~1 694	63
835~854	**75**	1 695~1 714	59
855~874	57	1 715~1 734	48
875~894	51	1 735~1 754	61
895~914	49	1 755~1 774	65
915~934	54	1 775~1 794	49
935~954	53	1 795~1 814	51
955~974	50	1 815~1 834	49
975~994	47	1 835~1 854	49
995~1 014	54	1 855~1 874	51
1 015~1 034	53	1 875~1 894	63
1 035~1 054	54	1 895~1 914	60
1 055~1 074	53	1 915~1 934	50
1 075~1 094	52	1 935~1 954	50
1 095~1 114	56	1 955~1 974	50

表3-7 基于归一化后的拉曼光谱和支持向量机判别算法的乳制品判别结果
（拉曼波段人工筛选）

光谱范围/cm^{-1}	识别率/%	光谱范围/cm^{-1}	识别率/%
350～405	60	**1 155～1 185**	**70**
485～540	**72**	1 185～1 230	56
590～670	66	1 230～1 300	65
780～820	61	**1 300～1 415**	**71**
820～915	**83**	**1 415～1 520**	**73**
915～985	62	1 550～1 580	61
985～1 030	58	1 580～1 605	59
1 030～1 060	61	1 605～1 640	65
1 060～1 115	61	1 640～1 730	63
1 115～1 155	68	1 730～1 800	60

3. 基于拉曼光谱特征提取统计融合的乳制品判别

已有文献中报道的数据融合方法主要包括数据层融合、特征层融合和决策层融合，有望提高数据利用效率，提高模型判别的准确性[①]。直接数据融合往往包含冗余信息，决策层融合需要多个分类器，操作相对复杂，数据的化学信息往往是模糊的，在一定程度上限制了其实际应用。相比较而言，基于清晰化学特征提取的数据融合策略简单、直观且实用[②③]。前述研究表明，乳制品拉曼光谱的不同特征区间对样品类别的判别方法有不同的贡献。归一化后，支持向量机分类算法的识别准确率得到提高，同时，一个以上光谱区间的准确率大于70%。结果显示出，通过特征谱区间的融合，可望进一步提高分类算法的识别准确率。测试条件包括：对所有光谱数据进行归一化，然后随机选择2/3的光谱数据作为训练集，其余1/3的光谱数据用作验证集，进行10次随机测试以获得平均准确率。如表3-8

---

① Hao N, Ping J C, Wang X, et al. Data fusion of near-infrared and mid-infrared spectroscopy for rapid origin identification and quality evaluation of Lonicerae japonicae flos[J]. Spectrochimica Acta Part A, Molecular and Biomolecular Spectroscopy, 2024, 320: 124590.

② Zhou L, Zhang C, Qiu Z J, et al. Information fusion of emerging non-destructive analytical techniques for food quality authentication: A survey[J]. TrAC Trends in Analytical Chemistry, 2020, 127: 115901.

③ Sha M, Zhang Z Y, Gui D D, et al. Data Fusion of ion Mobility Spectrometry Combined with Hierarchical Clustering Analysis for the Quality Assessment of Apple Essence[J]. Food Analytical Methods, 2017, 10(10): 3415-3423.

所示，可以清楚地看出，通过不同的组合融合，该算法的平均识别率高于基于相应单个特征谱区间的平均识别率。结果表明，最佳光谱特征融合区间包括 335～354 cm^{-1}、435～454 cm^{-1}、485～540 cm^{-1}、820～915 cm^{-1}、1 155～1 185 cm^{-1}、1 300～1 414 cm^{-1} 和 1 415～1 520 cm^{-1}，平均识别准确率可以从 58%（基于未归一化的全光谱数据）、91%（基于归一化的全谱数据）提高到 93%，并节省运行时间（从约 820 s 提高到 208 s）。基于特征区间融合的最佳分类结果可以达到 100%，如图 3-20 所示。

表 3-8 基于特征拉曼光谱和支持向量机识别算法融合的乳制品识别结果

拉曼光谱的范围/cm^{-1}	平均准确率/%	运行时间/s
335～354,435～454	76	30
335～354,435～454,835～854	84	38
485～540,820～915	84	78
485～540,820～915,1 115～1 185	85	90
485～540,820～915,1 115～1 185,1 300～1 415	88	143
485～540,820～915,1 115～1 185,1 300～1 414,1 415～1 520	89	192
485～540,820～915,1 115～1 185,1 415～1 520	87	138
335～354,435～454,485～540,835～854	83	61
335～354,435～454,835～854,1 115～1 185	83	51
335～354,435～454,835～854,1 115～1 185,1 300～1 415	90	102
335～354,435～454,835～854,1 115～1 185,1 300～1 414,1 415～1 520	90	151
335～354,435～454,485～540,835～854,1 115～1 185,1 300～1 414,1 415～1 520	91	174
335～354,435～454,485～540,820～915,1 115～1 185,1 300～1 414,1 415～1 520	93	208

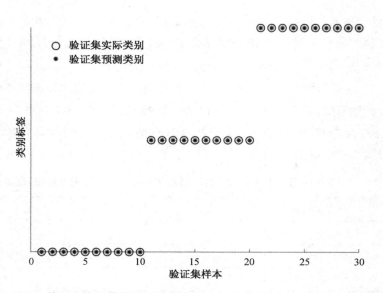

图 3-20　基于支持向量机的不同品牌 P1(a)、P2(b)和 P3(c)乳制品分类结果

3.2.3　小结

在本节研究工作中,提出了一种新的、简便的乳制品特征提取和融合识别研究策略。通过拉曼光谱和支持向量机算法的综合应用,可以清晰有效地提取拉曼光谱的特征区间,实验乳制品判别分析的最佳光谱特征区间为 335～354 cm^{-1}、435～454 cm^{-1}、485～540 cm^{-1}、820～915 cm^{-1}、1 155～1 185 cm^{-1}、1 300～1 414 cm^{-1} 和 1 415～1 520 cm^{-1}。通过归一化处理和特征融合可以进一步提高识别准确率,识别准确率从 58%(基于未归一化的全光谱数据)、91%(基于归一化的全谱数据)提高到 93%。支持向量机算法的总运行时间从约 820 s 大大缩短到 208 s。该方案具有样品信号采集简单、分析速度快、设备便携等优点。该方法在食品防伪和质量控制领域具有潜在的应用前景。

3.3　基于机器学习算法和拉曼光谱的乳制品鉴别特征分析

作为食品的重要组成部分,乳制品的质量安全风险可分为两类。一类是有害物质,包括非法添加剂、重金属、有害毒素、农药、兽药、抗生素残留等[1][2]。另一类是假冒伪劣产品,其检测指标可能都符合国家标准要求,但违法者利用不同品牌或产地之间的价格差异进行牟利[3]。针对这两种不同形式的风险,目前有多种检测和控制策略可供选择。对于具有明确特征的分子,如三聚氰胺、青霉素、阿莫西林等,可采用色谱-质谱法或表面增强拉曼光谱法等成分分析方法来识别目标分子并进行定量分析[4][5][6],该策略主要关注点是鉴定乳制品样品中的目标分子。Wang 等研究了山羊乳和牛乳制品低聚糖图谱的差异,检测了 27 种低聚糖,并使用主成分分析进行样品分类,确定乳糖-N-三糖是区分山羊乳和牛乳的潜在生物标志物[7]。Huang 等利用激光诱导击穿光谱结合线性判别分析、k-近邻、随机森林、支持向量机和卷积神经网络对奶粉中外源性蛋白质掺假的识别进行了研究。卷积神经网络模型表现出良好的性能,平均识别准确率为 97.8%[8]。Yang 等开发了一种基于磁性基底的表面增强拉曼光谱,并结合机器学习算法(k-近邻、支持向量机、决策树),以实现乳制品中喹诺酮类抗生素的超痕量检测[9]。对

---

[1] Ranveer S A, Harshitha C G, Dasriya V, et al. Assessment of developed paper strip based sensor with pesticide residues in different dairy environmental samples[J]. Current Research in Food Science, 2022, 6: 100416.

[2] Shan J R, Shi L H, Li Y C, et al. SERS-based immunoassay for amplified detection of food hazards: Recent advances and future trends[J]. Trends in Food Science & Technology, 2023, 140: 104149.

[3] Pan W, Liu W J, Huang X J. Rapid identification of the geographical origin of Baimudan tea using a Multi-AdaBoost model integrated with Raman Spectroscopy[J]. Current Research in Food Science, 2024, 8: 100654.

[4] Shishov A, Nizov E, Bulatov A. Microextraction of melamine from dairy products by thymol-nonanoic acid deep eutectic solvent for high-performance liquid chromatography-ultraviolet determination[J]. Journal of Food Composition and Analysis, 2023, 116: 105083.

[5] 苏心悦,马艳莉,翟晨,等. 表面增强拉曼光谱技术在液体食品品质安全检测中的研究进展[J]. 光谱学与光谱分析,2023,43(9):2657-2666.

[6] 李海闽,梁琪,陈卫平,等. 牛奶中阿莫西林含量表面增强拉曼光谱检测方法的建立[J]. 食品与机械,2019,35(2):87-91.

[7] Wang H Y, Zhang X Y, Yao Y, et al. Oligosaccharide profiles as potential biomarkers for detecting adulteration of caprine dairy products with bovine dairy products[J]. Food Chemistry, 2024, 443: 138551.

[8] Huang W H, Guo L B, Kou W P, et al. Identification of adulterated milk powder based on convolutional neural network and laser-induced breakdown spectroscopy[J]. Microchemical Journal, 2022, 176: 107190.

[9] Yang Z C, Chen G Q, Ma C Q, et al. Magnetic Fe_3O_4@COF@Ag SERS substrate combined with machine learning algorithms for detection of three quinolone antibiotics: Ciprofloxacin, norfloxacin and levofloxacin[J]. Talanta, 2023, 263: 124725.

于没有目标分子的类似样品以及成分复杂多样的乳制品样品,可以使用整体分析鉴别技术[1],机器学习算法与光谱表征的结合已成为该领域最重要的研发方向之一。例如,Singh等系统总结了机器学习模型在乳制品巴氏灭菌过程中的综合评价应用,指出机器学习方法可以有效地实时监测巴氏灭菌过程,预测潜在的设备故障,提高工艺效率,评估巴氏灭菌产品的质量[2]。

机器学习算法可以有效利用乳制品的特征数据,快速获得鉴别结果。例如,Cui等利用快速蒸发电离质谱与机器学习算法(判别分析、决策树、支持向量机和神经网络分类器)相结合,对乳脂奶油和非乳奶油的判别进行了研究,通过超参数优化和特征工程,识别准确率达到98.4%~99.6%[3]。Feng等将拉曼光谱与轻量梯度提升机、支持向量机、随机森林和极限梯度提升相结合,研究表明,在单一算法条件下,乳制品品牌分类的准确率超过90%,当这些算法结合使用时,准确率可达99%[4]。但这类方法也存在一些亟待改进的问题,如光谱特征分析[5][6]。对于机器学习判别算法而言,它们有时存在类似于"黑箱"问题,可解释性有限,因此,有必要深入研究光谱数据的特征区间是否会对相关算法产生影响,以及如何进一步提高效率,进而增强人们对此类问题的理解[7]。Pu等总结了常用的高光谱成像特征构建方法,并指出仍有更多的光谱信息分析方法有待不断探索[8]。拉曼光谱作为表征分子振动信息的前沿技术,在乳制品分析领域受到了广泛关注,显

---

[1] Silva M G, de Paula I L, Stephani R, et al. Raman spectroscopy in the quality analysis of dairy products: A literature review[J]. Journal of Raman Spectroscopy, 2021, 52(12): 2444-2478.

[2] Singh P, Pandey S, Manik S. A comprehensive review of the dairy pasteurization process using machine learning models[J]. Food Control, 2024, 164: 110574.

[3] Cui Y W, Lu W B, Xue J, et al. Machine learning-guided REIMS pattern recognition of non-dairy cream, milk fat cream and whipping cream for fraudulence identification[J]. Food Chemistry, 2023, 429: 136986.

[4] Feng Z K, Liu D, Gu J Y, et al. Raman spectroscopy and fusion machine learning algorithm: A novel approach to identify dairy fraud[J]. Journal of Food Composition and Analysis, 2024, 129: 106090.

[5] Ji H Z, Pu D D, Yan W J, et al. Recent advances and application of machine learning in food flavor prediction and regulation[J]. Trends in Food Science & Technology, 2023, 138: 738-751.

[6] Xue X, Sun H Y, Yang M J, et al. Advances in the application of artificial intelligence-based spectral data interpretation: A perspective[J]. Analytical Chemistry, 2023, 95(37): 13733-13745.

[7] Martini G, Bracci A, Riches L, et al. Machine learning can guide food security efforts when primary data are not available[J]. Nature Food, 2022, 3: 716-728.

[8] Pu H B, Yu J X, Sun D W, et al. Feature construction methods for processing and analysing spectral images and their applications in food quality inspection[J]. Trends in Food Science & Technology, 2023, 138: 726-737.

示出强大的应用潜力[1]。例如,Hussain Khan等将拉曼光谱与偏最小二乘回归模型相结合,证明了其在生乳在线监测方面的潜力[2]。

因此,本节围绕基于机器学习算法的光谱特征分析开展了研究。其中,支持向量机是一种优秀的高维数据分类器;极限学习机是一种单隐层前馈神经网络算法,具有较高的学习效率和泛化能力;卷积神经网络是一种新型的神经网络算法,具有较强的特征提取能力和优秀的泛化能力。故本节通过将这三种各具特色的机器学习算法与拉曼光谱相结合,研究和探索拉曼光谱特征与算法之间的关系[3]。

3.3.1 实验部分

1. 样品和设备

实验用乳制品均购置于南京苏果超市,其中,鼎鑫奶片标记为品牌P1,朴珍奶片标记为品牌P2,雪原奶片标记为品牌P3。每个品牌有40个样品。

使用便携式激光拉曼光谱仪采集乳制品样品的拉曼光谱信号,光谱仪型号:Prott-ezRaman-D3,厂家:美国恩威光电公司(Enwave Optronics)。使用仪器内置的控制软件进行光谱信号基线校正。激光器的激光波长为785 nm,激光功率约为450 mW,电荷耦合器件温度为-85℃,激光照射时间为50 s,光谱分辨率为1 cm^{-1},光谱范围为250~2 339 cm^{-1}。样品粉末化后直接上样采集,在实验过程中无需任何额外化学处理。

2. 数据处理

实验中涉及的支持向量机、极限学习机、卷积神经网络算法、归一化和欧氏距离计算均使用MATLAB平台(美国MathWorks公司)实现。

相关算法简介如下:

支持向量机算法如3.2节中相关介绍[4],首先假设一个训练集(T),

---

[1] Wang K Q, Li Z L, Li J J, et al. Raman spectroscopic techniques for nondestructive analysis of agri-foods: A state-of-the-art review[J]. Trends in Food Science & Technology, 2021, 118: 490-504.

[2] Hussain Khan H M, McCarthy U, Esmonde-White K, et al. Potential of Raman spectroscopy for in-line measurement of raw milk composition[J]. Food Control, 2023, 152: 109862.

[3] Li J X, Qing C C, Wang X Q, et al. Discriminative feature analysis of dairy products based on machine learning algorithms and Raman spectroscopy[J]. Current Research in Food Science, 2024, 8: 100782.

[4] Ni X F, Jiang Y R, Zhang Y S, et al. Identification of liquid milk adulteration using Raman spectroscopy combined with lactose indexed screening and support vector machine[J]. International Dairy Journal, 2023, 146: 105751.

$$T=\{(x_1,y_2),\cdots,(x_m,y_m)\}\in (X\times Y)^m,$$

其中 $x_i \in X = \mathbf{R}^n, y_i \in Y = \{1,-1\}(i=1,2,\cdots,m)$，$x_i$ 为拉曼光谱数据，y_i 为乳制品的品牌类别标签。选择合适的核函数 $K(x,x')$ 和参数 C，构造并求解优化问题：

$$\min_\alpha \frac{1}{2}\sum_{i=1}^{j}\sum_{j=1}^{m} y_i y_j \alpha_i \alpha_j K(x_i,x_j) - \sum_{j=1}^{m} \alpha_j$$

$$s.t. \sum_{i=1}^{m} y_i \alpha_i = 0, \quad 0 \leqslant \alpha_i \leqslant C, \quad i=1,\cdots,m$$

得到最优解 $\alpha^* = (\alpha_1^*,\cdots,\alpha_m^*)^T$。

本节采用径向基函数作为核函数，选择 α^* 的正分量 $(0 < \alpha_j^* < C)$ 并计算阈值 $b^* = y_j - \sum_{i=1}^{m} y_i \alpha_i^* K(x_i - x_j)$。构建决策函数来计算实验样本的品牌识别结果

$$f(x) = \text{sgn}(\sum_{i=1}^{m}\alpha_i^* y_i K(x,x_i) + b^*)。$$

对于极限学习机算法[①]，假设有 m 个实验样本 (x_i,y_i)，其中 $\boldsymbol{x}_i = [x_{i1},x_{i2},\cdots,x_{in}]^T \in \mathbf{R}^n, \boldsymbol{y}_i = [y_{i1},y_{i2},\cdots,y_{im}]^T \in \mathbf{R}^m$，对于具有 L 个隐层节点的单个隐层神经网络，存在

$$\sum_{f=1}^{L} \boldsymbol{\beta}_f g(\boldsymbol{\lambda}_f \cdot \boldsymbol{x}_i + b_f) = \boldsymbol{o}_i$$

其中 $i=1,2,\cdots,n$，$g(\boldsymbol{\lambda}_f \cdot \boldsymbol{x}_i + b_f)$ 是激活函数，$\boldsymbol{\lambda}_f = [\lambda_{f1},\lambda_{f2},\cdots,\lambda_{fn}]^T$ 是 f 的隐藏层单元的输出权重，b_f 是 f 隐藏层单位的偏置，$\boldsymbol{\beta}_f = [\beta_{f1},\beta_{f2},\cdots,\beta_{fm}]^T$ 是 f 层的输出权重。$\boldsymbol{\lambda}_f \cdot \boldsymbol{x}_i$ 表示 $\boldsymbol{\lambda}_f$ 和 $\boldsymbol{x}_i$ 的内积。

神经网络的优化目标是最大限度地减少输出误差，具体表现为：

$$\sum_{f=1}^{L} \parallel \boldsymbol{o}_i - \boldsymbol{y}_i \parallel = 0$$

存在 $\boldsymbol{\lambda}_f, \boldsymbol{x}_i$ 和 b_f，使得 $\sum_{f=1}^{L} \boldsymbol{\beta}_f g(\boldsymbol{\lambda}_f \cdot \boldsymbol{x}_i + b_f) = \boldsymbol{W}_i$。

此外，$\boldsymbol{H} \cdot \boldsymbol{\beta} = \boldsymbol{W}$，$\boldsymbol{H}$ 是隐藏层节点的输出，$\boldsymbol{\beta}$ 是输出权重，$\boldsymbol{W}$ 是预期输出。

$$\boldsymbol{H} = \begin{bmatrix} g(\lambda_1 \cdot x_1 + b_1) & \cdots & g(\lambda_L \cdot x_1 + b_L) \\ \vdots & \cdots & \vdots \\ g(\lambda_1 \cdot x_n + b_1) & \cdots & g(\lambda_L \cdot x_n + b_L) \end{bmatrix}_{n\times L}$$

---

① Song S, Wang Q Y, Zou X, et al. High-precision prediction of blood glucose concentration utilizing Fourier transform Raman spectroscopy and an ensemble machine learning algorithm[J]. Spectrochimica Acta Part A，Molecular and Biomolecular Spectroscopy，2023，303：123176.

$$\boldsymbol{\beta} = \begin{bmatrix} \beta_1^T \\ \vdots \\ \beta_L^T \end{bmatrix}_{L \times m} \quad \text{和} \quad \boldsymbol{W} = \begin{bmatrix} y_1^T \\ \vdots \\ y_n^T \end{bmatrix}_{n \times m}$$

其中，m 是输出数，H 是隐藏层输出矩阵，W 为训练集目标矩阵。

对于卷积神经网络算法[1][2]，设置每个研究对象类别有 m 个观测样本，其中 X 为输入的光谱数据，Y 为输出数据，用下式表示，其中 c 为类别：

$$X = \{x_{i,1}, x_{i,2}, \cdots, x_{i,n}\}_{i=1}^m, Y = \{y_{i,1}, y_{i,2}, \cdots, y_{i,c}\}_{i=1}^m$$

卷积层从输入样本中学习特征。在输入样本与卷积核之间进行卷积运算，偏移卷积结果，并使用激活函数进行非线性变换。其计算方法如下：

$$x_k^r = f\left(\sum_{i \in R_k} x_i^{r-1} * \omega_{i,k}^r + b_k^r\right)$$

式中，r 是图层的序号，x_k^r 是 r 层中 k 的特征输出，x_i^{r-1} 是 $r-1$ 层的输出和 r 层的输入，$\omega_{i,k}^r$ 是第 i 层的卷积滤波器，b_k^r 是偏差，R_k 是输入特征映射的集合。$f(\cdot)$ 是激活函数。

对于归一化，其计算公式为：

$$z = \frac{(z_{\max} - z_{\min}) \times (x - x_{\min})}{x_{\max} - x_{\min}} + z_{\min}$$

其中，x 是需要归一化处理的拉曼光谱数据，$x_{\min}$、$x_{\max}$ 和 $z_{\min}$、$z_{\max}$ 分别为原始数据和归一化数据区间的下限和上限。由于本节后续研究工作是将光谱数据归一化为 $[-1, 1]$，因此 $z_{\min} = -1$，$z_{\max} = 1$。

对于欧氏距离运算，其公式为 $d(u, v) = \sqrt{\sum_{i=1}^n (u_i - v_i)^2}$。$d$ 是两个样本 u 和 v 的拉曼光谱数据之间的欧氏距离。

3.3.2 结果和讨论

1. 基于拉曼光谱结合不同机器学习算法的乳制品判别分析

乳制品的拉曼光谱可以表征样品丰富的成分信息，具有快速、无损、便于数据采集等优点。如图 3-21 所示，实验采集得到乳制品的拉曼光谱峰相对尖锐，主

---

[1] Lu B X, Tian F, Chen C, et al. Identification of Chinese red wine origins based on Raman spectroscopy and deep learning[J]. Spectrochimica Acta Part A, Molecular and Biomolecular Spectroscopy, 2023, 291: 122355.

[2] Rong D, Wang H Y, Ying Y B, et al. Peach variety detection using VIS-NIR spectroscopy and deep learning[J]. Computers and Electronics in Agriculture, 2020, 175: 105553.

要来源于蛋白质、脂肪和碳水化合物[1],每个峰的可能归属见表3-9,不同品牌的乳制品的光谱具有很高的相似性,其峰值位置非常接近,可以将其作为重要的光谱表征数据,进一步与机器学习算法相结合来研究相似样本的分类。

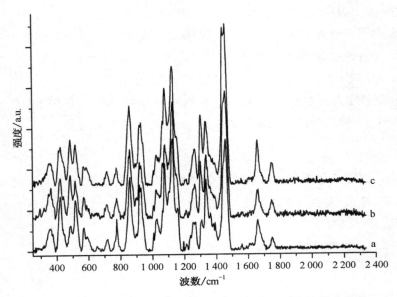

图3-21 不同品牌P1(a)、P2(b)和P3(c)乳制品的拉曼光谱图

将整个拉曼光谱直接依次导入到支持向量机、极限学习机、卷积神经网络算法中。随机选取70%的乳制品样本数据作为训练集,其余30%的样本数据作为测试集,执行10次运算,获得平均识别准确率。全光谱与支持向量机、极限学习机、卷积神经网络相结合的运算结果分别为33.3%、92.5%和100%,三种算法所需的时间也各不相同,大约分别消耗2 000 s、1 s和700 s。结果表明,将原始全光谱直接与支持向量机算法相结合无法实现对乳制品的有效鉴别,根据已有相关报道和研究小组的前期工作,支持向量机算法在处理类似问题时,需要进行光谱预处理操作,方能有望实现此类问题的有效判别[2]。极限学习机算法具有计算速度快和识别准确度高等方面的明显优势,卷积神经网络算法具有特征提取能力,呈

---

[1] Huang W, Fan D S, Li W F, et al. Rapid evaluation of milk acidity and identification of milk adulteration by Raman spectroscopy combined with chemometrics analysis[J]. Vibrational Spectroscopy, 2022, 123: 103440.

[2] Wang J F, Lin T H, Ma S Y, et al. The qualitative and quantitative analysis of industrial paraffin contamination levels in rice using spectral pretreatment combined with machine learning models[J]. Journal of Food Composition and Analysis, 2023, 121: 105430.

现出最高的识别准确率和适中的计算时间[①]。

表 3-9 不同品牌乳制品拉曼光谱主要峰值的可能归属分析

P1 品牌 波数/cm^{-1}	P2 品牌 波数/cm^{-1}	P3 品牌 波数/cm^{-1}	归属	可能的组分归属
362	365	363	乳糖	乳糖
423	425	426	葡萄糖	葡萄糖
487	487	486	$\delta(C—C—C)+\tau(C—O)$	碳水化合物
518	516	519	葡萄糖	葡萄糖
570	570	568	$\delta(C—C—O)+\tau(C—O)$	碳水化合物
719	716	717	$\nu(C—S)$	蛋白质
776	776	775	$\nu(C—C—O)$	碳水化合物
861	860	859	$\delta(C—C—H)+\delta(C—O—C)$	碳水化合物
920	921	921	$\delta(C—O—C)+\delta(C—O—H)+\nu(C—O)$	碳水化合物
1 007	1 006	1 008	苯环呼吸振动(苯丙氨酸), $\nu(C—C)_{环}$	蛋白质
1 029	1 025	1 026	$\nu(C—O)+\nu(C—C)+\delta(C—O—H)$	碳水化合物
1 078	1 080	1 080	$\nu(C—O)+\nu(C—C)+\delta(C—O—H)$	碳水化合物
1 129	1 130	1 128	$\nu(C—O)+\nu(C—C)+\delta(C—O—H)$	碳水化合物
1 270	1 266	1 267	$\gamma(CH_2)$	碳水化合物
1 310	1 303	1 305	$\tau(CH_2)$	脂肪
1 337	1 338	1 336	$\nu(C—O)+\delta(C—O—H)$	碳水化合物
1 461	1 460	1 461	$\delta(CH_2)$	脂肪、碳水化合物
1 660	1 664	1 660	$\nu(C=O)$酰胺Ⅰ, $\nu(C=C)$	脂肪、蛋白质
1 751	1 747	1 750	$\nu(C=O)_{酯}$	脂肪

注: ν:伸缩振动, δ:变形振动, τ:扭曲振动, γ:面外弯曲振动。

---

① Chen Z F, Khaireddin Y, Swan A K. Identifying the charge density and dielectric environment of graphene using Raman spectroscopy and deep learning[J]. Analyst, 2022, 147(9): 1824-1832.

2. 基于特征谱结合不同机器学习算法的乳制品鉴别分析

归一化处理有望消除光谱数据量纲的影响,提高分类器的判别准确率。研究将拉曼光谱数据归一化为 -1 到 1 的范围,然后将数据分别导入三种机器学习算法中。计算结果表明,支持向量机算法的识别准确率提高到 100%,极限学习机算法的识别准确率提高到 94.4%,卷积神经网络算法的识别准确率保持在 100%。运算时间分别约为 1 750 s、1 s 和 750 s。结果表明,实验数据的归一化处理对提高不同机器学习算法的识别准确率有很好的效果。

随后,本节通过将不同的光谱特征峰与不同的机器学习算法相结合,研究了识别准确率的变化情况,如表 3-10 所示。结果清楚地表明,不同特征波段对分类算法有不同的贡献,同时,同一特征波段对应不同分类算法的识别准确率也有一定的差异。对于支持向量机算法,拉曼特征区间内识别准确率排名前三位的波段分别是 1 410~1 500 cm^{-1}、890~980 cm^{-1} 和 1 100~1 180 cm^{-1}。对于极限学习机算法,拉曼特征区间的前三名分别是 1 410~1 500 cm^{-1}、1 050~1 100 cm^{-1} 和 810~890 cm^{-1}。对于卷积神经网络算法,前三个拉曼特征区间分别为 1 410~1 500 cm^{-1}、890~980 cm^{-1} 和 810~890 cm^{-1}。

表 3-10 不同拉曼光谱特征区间与三种机器学习算法相结合的判别结果

光谱特征区间/cm^{-1}	支持向量机(识别率/%)	极限学习机(识别率/%)	卷积神经网络(识别率/%)
280~390	94.4	87.2	98.3
390~460	89.2	84.2	93.3
460~500	81.1	71.1	96.9
500~550	86.4	71.9	90.3
550~630	97.8	91.1	98.3
630~690	84.2	71.9	84.2
690~745	66.7	52.5	66.8
745~810	85.3	64.7	85.0
810~890	98.9	96.4	98.6
890~980	99.7	93.1	98.9
980~1 015	76.9	63.1	89.2
1 015~1 050	83.1	78.6	92.2

续表 3 - 10

光谱特征区间/cm^{-1}	支持向量机（识别率/%）	极限学习机（识别率/%）	卷积神经网络（识别率/%）
1 050～1 100	99.2	98.6	98.3
1 100～1 180	99.4	94.7	96.7
1 180～1 225	76.4	64.7	76.4
1 225～1 290	92.2	87.5	95.0
1 290～1 320	98.3	93.3	95.8
1 320～1 410	98.6	91.1	96.7
1 410～1 500	100	99.4	99.7
1 500～1 630	84.7	70.3	93.6
1 630～1 710	99.2	89.4	98.3
1 710～1 780	83.9	63.6	96.1
1 780～2 339	68.9	59.7	90.8

识别准确率高的特征区间还能为我们提供一些有关乳制品分类差异的物质信息。基于支持向量机算法识别准确率较高的前 9 个光谱特征区间如图 3 - 22 所示，相应拉曼光谱峰归属分析如下。550～630 cm^{-1} 波段主要来源于糖苷环骨架变形振动，810～890 cm^{-1}、890～980 cm^{-1}、1 050～1 100 cm^{-1} 和 1 100～1 180 cm^{-1} 波段主要来源于糖苷键。1 290～1 320 cm^{-1} 波段主要来自脂类，1 320～1 410 cm^{-1} 波段主要来自碳水化合物，1 410～1 500 cm^{-1} 波段主要来自脂肪和碳水化合物。1 630～1 710 cm^{-1} 波段主要来自不饱和脂肪酸的 C=C 伸缩振动和蛋白质酰胺Ⅰ键 CONH 中的 C=O 伸缩振动[1]。也有一些光谱特征区域在与算法结合时表现出较低的识别准确率，如 690～745 cm^{-1} 和 980～1 015 cm^{-1}（可能源自蛋白质分子）以及 1 780～2 339 cm^{-1}，这一区域可主要归属于光谱噪声，没有明显的光谱峰。

---

[1] Almeida M R, de S Oliveira K, Stephani R, et al. Fourier-transform Raman analysis of milk powder: A potential method for rapid quality screening[J]. Journal of Raman Spectroscopy, 2011, 42(7): 1548 - 1552.

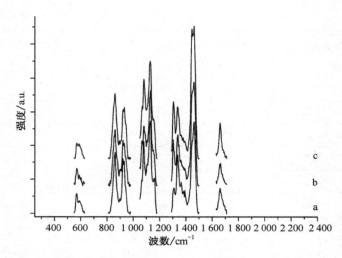

图 3-22 不同品牌 P1(a)、P2(b) 和 P3(c) 乳制品的拉曼光谱特征区间

此外，由于特征波段只是原始数据的一部分，使得每个分类器的运算时间大大缩短。除少数特征波段（280～390 cm^{-1}，1 500～1 630 cm^{-1}，1 780～2 339 cm^{-1}）的运算时间超过 100 s 外，其他波段在与支持向量机算法结合中的运算时间均在 100 s 以内。极限学习机算法运行时间仅为 0.3 s，卷积神经网络算法运行时间仅为 50 s。

在结合单个谱峰特征区间和上述算法的分析的基础上，进一步研究了多光谱区间融合对各算法识别准确率的影响情况。实验将各算法识别准确率对应的谱区间依次与基于支持向量机算法识别准确率对应的谱特征区间前 3 个波段、前 9 个波段进行融合分析，识别准确率的变化情况如表 3-11 所示。可以看出，不同机器学习算法在某些融合波段都取得了改善的判别结果。例如，支持向量机算法在 890～980 cm^{-1}、1 410～1 500 cm^{-1} 融合光谱特征波段内识别准确率为 100%，耗时约 200 s。极限学习机算法在 890～980 cm^{-1}、1 410～1 500 cm^{-1} 融合光谱特征波段内识别准确率也为 100%，并且耗时小于 0.3 s。卷积神经网络算法在 890～980 cm^{-1}、1 050～1 180 cm^{-1}、1 410～1 500 cm^{-1} 的融合光谱特征波段内的识别准确率为 100%，耗时约 80 s。基于卷积神经网络算法的最优识别结果如图 3-23 所示。这一研究结果表明，通过合理的拉曼光谱特征区间融合，并结合相应的机器学习算法，可以显著提高识别准确率，并且与全光谱数据相比，能够保持较高的运算效率。

表 3-11　不同拉曼光谱特征融合区间与三种机器学习算法相结合的判别结果

光谱特征区间/cm^{-1}	支持向量机（识别率/%）	极限学习机（识别率/%）	卷积神经网络（识别率/%）
890～980，1 410～1 500	100	100	99.4
890～980，1 100～1 180，1 410～1 500	100	99.4	99.4
1 050～1 100，1 410～1 500	100	99.7	99.7
810～890，1 050～1 100，1 410～1 500	100	99.7	98.3
810～980，1 410～1 500	100	98.9	99.4
890～980，1 050～1 180，1 410～1 500	100	99.7	100
890～980，1 050～1 180，1 410～1 500，1 630～1 710	100	100	100
810～980，1 050～1 180，1 410～1 500，1 630～1 710	100	100	99.2
810～980，1 050～1 180，1 320～1 500，1 630～1 710	100	100	99.7
810～980，1 050～1 180，1 290～1 500，1 630～1 710	100	100	99.7
550～630，810～980，1 050～1 180，1 290～1 500，1 630～1 710	100	100	100

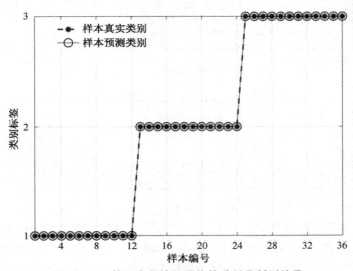

图 3-23　基于卷积神经网络的乳制品判别结果

（y 轴：标签 1 代表品牌 P1，标签 2 代表品牌 P2，标签 3 代表品牌 P3）

3. 光谱特征区间的进一步统计分析

为了进一步分析上述拉曼光谱特征的统计波动情况,进一步开展了使用基于拉曼光谱波段 $1\,410\sim1\,500\ cm^{-1}$ 的质量波动控制图分析。计算步骤如下:第一步是计算 P1 品牌乳制品的拉曼光谱均值,并将其作为该品牌光谱真实值的最佳估计值。第二步是计算 P1 品牌的各实验样品与该均值之间的欧氏距离,将此结果代入单值移动极差控制图运算公式,得到单值和移动极差各控制限,进而绘制出相应的控制图。第三步是计算 P2 品牌和 P3 品牌乳制品各个实验样品与 P1 品牌均值间的欧氏距离,并将结果绘制在控制图上,如图 3 - 24 所示。可以看出,基于对识别准确率贡献最大的光谱特征区间,P1 品牌的各个样本都存在质量波动,但均在中心线附近正常波动,处于可控范围内(图 3 - 24a 和图 3 - 24d)。在 P2 品牌的样本中,在单值控制图中出现有 11 个样本跃出了控制限(图 3 - 24b),在移动极差控制图中有 7 个跃出了控制限(图 3 - 24e)。在 P3 品牌中,单值控制图中的全部 40 个都跃出了控制限(图 3 - 24c),在移动极差控制图中有 15 个跃出了控制限(图 3 - 24f)。因此,这反映出三个品牌样品在该光谱特征区间内存在一定的差异,但也有 P2、P3 品牌的一定数量样品仍落入 P1 品牌的控制限内,研究结果直观地显示了实验样本的一定的统计波动情况。

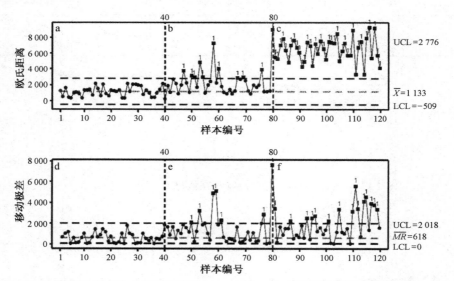

图 3 - 24 基于拉曼光谱特征区间($1\,410\sim1\,500\ cm^{-1}$)欧氏距离运算结果,品牌 P1(a)、品牌 P2(b)、品牌 P3(c)单值控制图和品牌 P1(d)、品牌 P2(e)、品牌 P3(f)移动极差控制图

注:UCL 表示控制上限,LCL 表示控制下限,$\bar{X}$ 表示单值控制图的平均值,$\overline{MR}$ 表示移动极差控制图的平均值。

随后,两个光谱特征区间 1 410～1 500 cm⁻¹ 和 890～980 cm⁻¹,以及光谱特征区间 1 100～1 180 cm⁻¹,被选择进行欧氏距离计算。具体来说,以 P1 品牌的拉曼光谱均值作为真值的最佳估计,分别在前面两个及三个光谱特征区间条件下,计算出 P1 品牌各个样本与均值之间的欧氏距离,以及 P2 品牌、P3 品牌与 P1 品牌均值之间的欧氏距离,依据运算结果可分别绘制出图 3-25 和图 3-26。可以

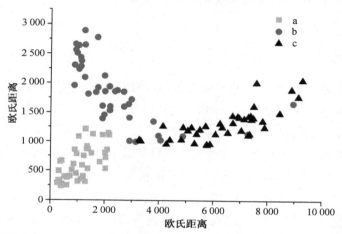

图 3-25 不同品牌 P1(a)、P2(b) 和 P3(c) 乳制品的基于拉曼光谱特征区间的欧氏距离二维图
(x 轴:1 410～1 500 cm⁻¹,y 轴:890～980 cm⁻¹)

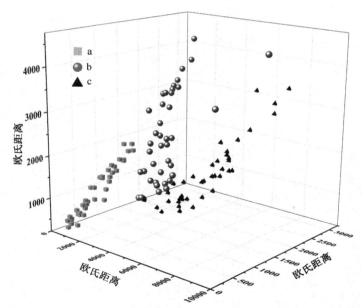

图 3-26 不同品牌 P1(a)、P2(b) 和 P3(c) 乳制品的基于拉曼光谱特征区间的欧氏距离三维图
(x 轴:1 410～1 500 cm⁻¹,y 轴:890～980 cm⁻¹,z 轴:1 100～1 180 cm⁻¹)

看出,在 1 410～1 500 cm^{-1} 和 890～980 cm^{-1} 的条件下,P1 品牌和 P2、P3 品牌在二维空间上存在一定程度的差异。在 1 410～1 500 cm^{-1}、890～980 cm^{-1} 和 1 100～1 180 cm^{-1} 三维空间的条件下,在 P1 品牌和 P2、P3 品牌各样品之间也有一定程度的差异,各样本间可进行一定程度的区分。这一结果显示出样本在特征光谱区间条件下的分布差异情况,揭示了机器学习算法有效识别的光谱特征基础。

3.3.3 小结

本节研究将支持向量机、极限学习机、卷积神经网络算法和拉曼光谱相结合,研究了乳制品鉴别的光谱特征作用,可得出如下结论:(1)不同的机器学习算法所对应的最优光谱特征区间存在差异;(2)少量的光谱特征区间融合可以在一定程度上提高机器学习算法的识别准确率和计算效率;(3)作为机器学习算法的数据输入部分,拉曼光谱不同的光谱特征区间有不同的贡献。通过统计分布分析,样本空间分布可以直观显示具有高分辨能力的光谱特征区间。因此,本节研究所建立的一整套特征光谱分析策略将有助于读者进一步理解不同类别乳制品判别的表征基础。

第 4 章　乳制品高维拉曼光谱数据的构建研究

前述的研究显示出适当的拉曼光谱预处理、特征提取操作在质量判别过程中可以起到重要的作用。不过,我们也注意到拉曼光谱数据维度整体处于二维状态,为此,本章将从数据构建视角,设计研究二维相关拉曼光谱,探讨论证乳制品高维拉曼光谱表征情境下质量变化的差异性情况。

4.1　激光微扰二维相关拉曼光谱用于乳制品质量判别

乳制品质量与安全评估一直是食品科学和质量监管领域的热门话题之一。常用的质量判别分析方法可简要分为以下三类:一是通过聘请专业人员,根据乳制品的颜色、味道、组织状态等,对乳制品的质量安全水平进行观察和评估,该类方法称为主观品鉴法。二是主要利用电子鼻、电子舌等仿生模拟仪器获取乳制品的嗅觉和味觉信息。该方法一般需结合主成分分析等数学模型,可以对乳制品的质量安全水平进行判别分析[1][2]。三是利用常规光谱法、色谱法和质谱法等,可以对乳制品的营养成分、药物残留、非法添加剂和微生物指标含量等进行分析[3][4][5]。与传统方法相比,拉曼光谱法是一种快速分析方法,具有样品无需预处理即可检

---

[1] Peris M, Escuder-Gilabert L. Electronic noses and tongues to assess food authenticity and adulteration[J]. Trends in Food Science & Technology, 2016, 58: 40 – 54.

[2] Yang C J, Ding W, Ma L J, et al. Discrimination and characterization of different intensities of goaty flavor in goat milk by means of an electronic nose[J]. Journal of Dairy Science, 2015, 98(1): 55 – 67.

[3] Poonia A, Jha A, Sharma R, et al. Detection of adulteration in milk: A review[J]. International Journal of Dairy Technology, 2017, 70(1): 23 – 42.

[4] Zhu W X, Yang J Z, Wang Z X, et al. Rapid determination of 88 veterinary drug residues in milk using automated TurborFlow online clean-up mode coupled to liquid chromatography-tandem mass spectrometry[J]. Talanta, 2016, 148: 401 – 411.

[5] Tolmacheva V V, Apyari V V, Furletov A A, et al. Facile synthesis of magnetic hypercrosslinked polystyrene and its application in the magnetic solid-phase extraction of sulfonamides from water and milk samples before their HPLC determination[J]. Talanta, 2016, 152: 203 – 210.

测、操作相对简单、检测时间短、灵敏度高等优点[1][2]。因此,拉曼光谱技术在乳制品检测领域得到了广泛关注,但相关研究案例主要集中于三聚氰胺、蔗糖等非法物质的检测[3][4][5]。

对于拉曼光谱数据分析,它可以用于乳制品的定性分析、定量分析和结构分析,常用的数据分析方法包括朗伯比尔定律等。然而,这种分析方法只使用最大拉曼散射峰值强度,数据利用率相对较低。近年来,Noda教授提出了一种二维相关性分析策略[6][7][8]。该策略的基本思想是收集目标系统(如乳制品)在外部扰动条件下的动态光谱数据(如拉曼光谱),然后使用二维相关分析算法对这些数据进行分析,最终获得被测系统的高维信息。这种新策略可以提供系统的更多光谱细节,并提高光谱分辨率。常用的扰动包括温度、浓度和化学反应等[9][10]。Lei 等报道,以温度为扰动,利用二维相关红外光谱法可以检测奶粉中的结晶乳糖[11]。然而,使用上述微扰方法可以提高光谱分辨率,同时也增加了实验时间消耗,此外,

---

[1] Ullah R, Khan S, Khan A, et al. Infant gender-based differentiation in concentration of milk fats using near infrared Raman spectroscopy[J]. Journal of Raman Spectroscopy, 2017, 48(3): 363 – 367.

[2] Chen Y L, Li X L, Yang M, et al. High sensitive detection of penicillin G residues in milk by surface-enhanced Raman scattering[J]. Talanta, 2017, 167: 236 – 241.

[3] Nieuwoudt M K, Holroyd S E, McGoverin C M, et al. Rapid, sensitive, and reproducible screening of liquid milk for adulterants using a portable Raman spectrometer and a simple, optimized sample well[J]. Journal of Dairy Science, 2016, 99(10): 7821 – 7831.

[4] Li X Y, Feng S L, Hu Y X, et al. Rapid detection of melamine in milk using immunological separation and surface enhanced Raman spectroscopy[J]. Journal of Food Science, 2015, 80(6): C1196 – C1201.

[5] Hu Y X, Lu X N. Rapid detection of melamine in tap water and milk using conjugated "one-step" molecularly imprinted polymers-surface enhanced Raman spectroscopic sensor[J]. Journal of Food Science, 2016, 81(5): N1272 – N1280.

[6] Noda I. Recent developments in two-dimensional (2D) correlation spectroscopy[J]. Chinese Chemical Letters, 2015, 26(2): 167 – 172.

[7] Noda I, Ozaki Y. Two-Dimensional Correlation Spectroscopy-Applications in Vibrational and Optical Spectroscopy[M]. Chichester: John Wiley & Sons, Inc., 2004: 1 – 195.

[8] Eads C D, Noda I. Generalized correlation NMR spectroscopy[J]. Journal of the American Chemical Society, 2002, 124(6): 1111 – 1118.

[9] Park Y, Noda I, Jung Y M. Novel developments and applications of two-dimensional correlation spectroscopy[J]. Journal of Molecular Structure, 2016, 1124: 11 – 28.

[10] Zhang Y L, Chen J B, Lei Y, et al. Discrimination of different red wine by Fourier-transform infrared and two-dimensional infrared correlation spectroscopy[J]. Journal of Molecular Structure, 2010, 974(1/2/3): 144 – 150.

[11] Lei Y, Zhou Q, Zhang Y L, et al. Analysis of crystallized lactose in milk powder by Fourier-transform infrared spectroscopy combined with two-dimensional correlation infrared spectroscopy[J]. Journal of Molecular Structure, 2010, 974(1/2/3): 88 – 93.

这种二维相关数据处理策略在乳制品相关研究中的实验案例还较少。

为此,本节研究了一种新的二维相关拉曼光谱构建方法,用于乳制品的判别分析。这一新的分析方法主要包括以下三个方面:第一,对于复杂的乳制品系统,传统的仪器分析方法可以测量样品的一些理化指标,但有时很难区分以次充好的样品。本节所开发的分析方法可用于表征和区分相似样品。第二,传统的仪器分析方法通常需要对样品进行预处理,而且耗时较长。二维相关分析方法可以提高光谱分辨率,但也增加了检测时间。本节工作采用拉曼光谱,无需预处理,测试速度快。使用激光作为微扰动时,可以快速收集样品的动态光谱,只需要几分钟的总分析测试时间[1]。第三,二维相关分析结果可以在三维空间中显示样本的特征,使人们能够实现对测试样本的直接识别。研究应用相关系数法对结果开展进一步分析,可以计算出每个样本之间的相关系数值,实现了定量描述各样本间的相似性[2]。

4.1.1 实验部分

1. 样品与测量

实验用乳制品均购置于南京苏果超市,其中,乃捷尔乳制品标记为品牌 P1,雪原乳制品标记为品牌 P2,贝因美乳制品标记为品牌 P3,飞鹤乳制品标记为品牌 P4,美素佳儿乳制品标记为品牌 P5,启赋乳制品标记为品牌 P6,雀巢乳制品标记为品牌 P7,伊利乳制品标记为品牌 P8。

将几毫克乳制品粉末分别装入自制多孔芯片的上样孔中。然后使用便携式激光拉曼光谱仪获得样品的拉曼光谱,光谱仪型号:Prott-ezRaman-D3,厂家:美国恩威光电公司(Enwave Optronics)。激光激发波长 785 nm,激光功率约 450 mW,激光光斑直径约 100 μm。照射时间从 10 s 变为 50 s,间隔为 10 s。光谱范围 250~2 010 cm^{-1},光谱分辨率为 1 cm^{-1}。

2. 数据分析

通过光谱仪自带的软件 SLSR Reader V8.3.9 对获得的拉曼光谱进行基线校正,然后使用 Savitzky Golay 算法对拉曼光谱进行平滑。二维相关拉曼光谱的

---

[1] Moros J, Javier Laserna J. Unveiling the identity of distant targets through advanced Raman-laser-induced breakdown spectroscopy data fusion strategies[J]. Talanta, 2015, 134: 627-639.

[2] Zhang Z Y, Sha M, Wang H Y. Laser perturbation two-dimensional correlation Raman spectroscopy for quality control of bovine colostrum products[J]. Journal of Raman Spectroscopy, 2017, 48(8): 1111-1115.

计算使用"2D shige"软件实现,该软件由日本 Kwansei Gakuin University(关西学院大学)Yukihiro Ozak 教授小组的 Shigeaki Morita 开发。在二维相关分析中,使用平均谱作为参考谱。使用 Excel 软件(美国微软公司,Microsoft Corporation)获得样品间的相关系数。

4.1.2 结果和讨论

1. 乳制品的拉曼光谱和二维相关拉曼光谱表征

实验首先采集了 P1 乳制品的拉曼光谱和构建了二维相关拉曼光谱。常规拉曼光谱如图 4-1 所示。根据已有的相关报道[1][2][3],主要振动谱带可尝试归属如下:1 674 cm^{-1} 处的谱带可归因于 C=O 伸缩振动和 C=C 伸缩振动,分别来自蛋白质酰胺 I 的 CONH 基团和不饱和脂肪酸[4][5],最大的拉曼光谱峰在 1 456 cm^{-1},其归属可能来自脂肪和糖分子的 CH_2 变形振动,1 350 cm^{-1} 处的谱带可能与糖类分子的 C—O 伸缩振动、C—O—H 变形振动有关,1 250 cm^{-1} 处的谱带可归因于糖类分子的 CH_2 扭曲振动和肽段基团的 N—H 弯曲振动[6]。1 128 cm^{-1} 和 1 090 cm^{-1} 处的谱带可归属于糖类分子的 C—O 伸缩振动、C—C 伸缩振动和 C—O—H 变形振动。857 cm^{-1} 处的谱带可归因于糖类分子的 C—O—C 变形振动和 C—O—H 变形振动。1 009 cm^{-1} 处的谱带是一个特殊的拉曼峰,主要与蛋白质苯丙氨酸的苯环呼吸振动有关。在 300～800 cm^{-1} 的范围内,有 10 多个谱带可归属于 C—C—O 变形、C—C—C 变形、C—O 扭曲振动和 C—S

---

[1] Rygula A, Majzner K, Marzec K M, et al. Raman spectroscopy of proteins: A review[J]. Journal of Raman Spectroscopy, 2013, 44(8): 1061-1076.

[2] Almeida M R, de S Oliveira K, Stephani R, et al. Fourier-transform Raman analysis of milk powder: A potential method for rapid quality screening[J]. Journal of Raman Spectroscopy, 2011, 42(7): 1548-1552.

[3] Mendes T O, Junqueira G M A, Porto B L S, et al. Vibrational spectroscopy for milk fat quantification: Line shape analysis of the Raman and infrared spectra[J]. Journal of Raman Spectroscopy, 2016, 47(6): 692-698.

[4] Moros J, Garrigues S, de la Guardia M. Evaluation of nutritional parameters in infant formulas and powdered milk by Raman spectroscopy[J]. Analytica Chimica Acta, 2007, 593(1): 30-38.

[5] Zhou Q, Sun S Q, Yu L, et al. Sequential changes of main components in different kinds of milk powders using two-dimensional infrared correlation analysis[J]. Journal of Molecular Structure, 2006, 799(1/2/3): 77-84.

[6] Dahlenborg H, Millqvist-Fureby A, Brandner B D, et al. Study of the porous structure of white chocolate by confocal Raman microscopy[J]. European Journal of Lipid Science and Technology, 2012, 114(8): 919-926.

拉伸振动等①。图 4-1 同时显示了 P1 乳制品在不同激光照射时间(10 s 至 50 s)下的拉曼光谱,图示显示出,随着激光照射时间的增加,拉曼光谱的信号强度逐渐增加,特别是在 300~800 cm^{-1} 范围内,拉曼光谱信号逐渐清晰。其原因可能在于随着激光源照射和信号收集时间的增加,有助于微弱拉曼散射信号的获取,同时,激光的热效应可能会导致测试样品的振动信号发生微小的变化。故可利用激光作为扰动,尝试建立一种乳制品的二维相关性分析系统。

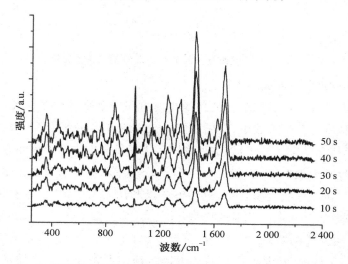

图 4-1　P1 品牌乳制品在不同激光照射时间(10 s 至 50 s)下的拉曼光谱

采用乳制品在不同激光照射时间下的拉曼光谱作为动态光谱,以平均谱作为参考谱,通过二维相关分析可以获得二维相关拉曼光谱,如图 4-2 所示。正相关强度和负相关强度由未出现阴影区域和有阴影区域表示。随着激光照射时间的增加,测试样品的拉曼信号增强,故相关信号主要为正信号。二维相关分析可以有效地在三维空间中提高原始光谱的分辨率,因此,其结果可以提供实验样品系统在外加扰动下变化的更多光谱细节②。测试样品在(1 670,1 670)、(1 450,1 450)、(1 330,1 330)、(1 250,1 250)、(1 105,1 105)、(860,860)周围有 6 个具有正强度的自动峰,在(1 670,1 450)、(1 450,1 670)等周围有 30 多个具有明显正强

---

① Rodrigues Júnior P H, de Sá Oliveira K, deAlmeida C E R, et al. FT-Raman and chemometric tools for rapid determination of quality parameters in milk powder: Classification of samples for the presence of lactose and fraud detection by addition of maltodextrin[J]. Food Chemistry, 2016, 196: 584-588.

② Yang R J, Liu R, Dong G M, et al. Two-dimensional hetero-spectral mid-infrared and near-infrared correlation spectroscopy for discrimination adulterated milk[J]. Spectrochimica Acta Part A, Molecular and Biomolecular Spectroscopy, 2016, 157: 50-54.

度的交叉峰。

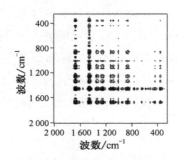

图 4-2 P1 品牌乳制品的同步二维相关拉曼光谱

2. 乳制品的相关系数分析

乳制品的拉曼光谱和二维相关拉曼光谱可用于表征其化学性质和分析其质量变化。在传统的分析方法中,常用来区分被测样品差异的策略可以简要分为两种:一种是直接视觉定性分析,但这种方法不能提供样本之间的定量差异信息,当样本具有高度相似性时,也不能区分样本。另一种是基于如朗伯比尔定律,结合最大特征谱峰进行定量分析,不过这种方法仅利用了谱峰值,信息利用率相对较低。此外,这种方法有时不能反映样本之间的总体光谱差异。相关系数分析是一种定量评估实验样品之间相似性的分析方法,它可以充分利用光谱的所有数据[1]。在本节工作中,考虑到乳制品在生产过程中通常存在质量波动,因此随机选择 6 个 P1 乳制品样品,分别基于拉曼光谱和二维相关拉曼光谱研究其相似性,如表 4-1 和表 4-2 所示。在表 4-1 乳制品之间的相关系数计算过程中使用的拉曼光谱激光照射时间为 50 s。从表 4-1 结果可以看出,不同样品之间的相关系数大于 0.985,表明 P1 乳制品之间具有较高的相似性。表 4-2 显示了 P1 乳制品基于其二维相关拉曼光谱的相关系数值。从该表中可以看出,相关系数大于 0.970,也表明各乳制品之间具有较高的相似性。通过比较表 4-1 和表 4-2,可以发现基于拉曼光谱获得的相关系数大于基于二维相关拉曼光谱计算的相关系数,这可能是因为二维相关拉曼光谱可以提高光谱分辨率,并进一步突出了样品细节的差异。

---

[1] Chen J B, Zhou Q, Noda I, et al. Quantitative classification of two-dimensional correlation spectra[J]. Applied Spectroscopy, 2009, 63(8): 920-925.

表 4-1 基于 P1 乳制品各样品间的拉曼光谱相关系数运算结果

	P1-1	P1-2	P1-3	P1-4	P1-5	P1-6
P1-1	1	0.992	0.990	0.987	0.992	0.987
P1-2		1	0.989	0.986	0.992	0.988
P1-3			1	0.989	0.991	0.990
P1-4				1	0.990	0.987
P1-5					1	0.989
P1-6						1

表 4-2 基于 P1 乳制品各样品间的二维相关拉曼光谱相关系数运算结果

	P1-1	P1-2	P1-3	P1-4	P1-5	P1-6
P1-1	1	0.982	0.982	0.973	0.983	0.979
P1-2		1	0.978	0.973	0.981	0.978
P1-3			1	0.976	0.981	0.978
P1-4				1	0.976	0.972
P1-5					1	0.976
P1-6						1

为了克服乳制品质量正常波动的影响,采用光谱平均值来估计其理论值。在这项工作中,还计算了实验各样品与其基于拉曼光谱和二维相关拉曼光谱的均值之间的相关系数,如表 4-3 所示。结果表明,基于拉曼光谱,乳制品与其均值之间的相关系数为 0.996±0.001,平均值为 0.996,标准偏差为 0.001。基于二维相关拉曼光谱,乳制品与其均值之间的相关系数位于 0.991±0.003,平均值为 0.991,标准偏差为 0.003。可见,乳制品之间的相似性确实较高。同时,基于拉曼光谱的相关系数也大于基于二维拉曼相关谱的相关系数,这可以归因于通过二维相关分析提高了样品的光谱分辨率。

表 4-3 分别基于不同 P1 乳制品的拉曼光谱和二维相关拉曼光谱相关系数运算结果

	P1-1	P1-2	P1-3	P1-4	P1-5	P1-6
P1(拉曼光谱,均值)	0.996	0.996	0.996	0.994	0.997	0.994
P1(二维相关拉曼光谱,均值)	0.992	0.992	0.991	0.986	0.994	0.989

3. 其他品牌乳制品的二维相关拉曼光谱表征及相关系数分析

其他品牌乳制品也是以牛乳为主要原料制备得来，随机选择市场上的六个品牌的乳制品，在相同的实验条件下收集它们的拉曼光谱，进一步考察样品间的相似性。实验也进行了二维相关分析，结果如图 4-3 所示，(A)到(F)依次对应的是品牌 P3 到 P8。结果显示出，不同品牌乳制品的二维相关拉曼光谱在三维空间中也表现出丰富的化学特征。例如，P3 乳制品有如(1 450, 1 450)等 6 个以上的自动峰，以及如(1 670, 1 450)等 30 多个交叉峰。从图 4-3 和图 4-2 的比较可以看出，各品牌乳制品之间存在一定的差异。

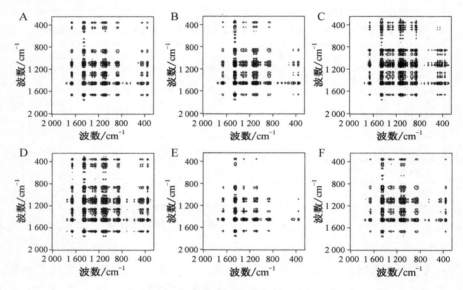

图 4-3　不同品牌乳制品的同步二维相关拉曼光谱

注：(A)到(F)依次为品牌 P3 到 P8。

随后，基于拉曼光谱和二维相关拉曼光谱，分别计算得到 P3 至 P8 品牌乳制品样品和 P1 品牌乳制品（均值）之间的相关系数，如表 4-4 所示。结果显示出，基于拉曼光谱，P3 至 P8 品牌乳制品和 P1 品牌乳制品（均值）间的相关系数在 0.765～0.830 之间，小于 0.996±0.001（表 4-3）。同时，基于二维相关拉曼光谱也显示出，P3 至 P8 品牌乳制品和 P1 品牌乳制品（均值）的相关系数在 0.683～0.761 之间，也小于 0.991±0.003（表 4-3）。因此，可以根据相关系数将 P3 至 P8 品牌乳制品样品与 P1 品牌乳制品区分开来。此外，从表 4-4 中的相关系数比较发现，基于二维相关拉曼光谱的运算结果 0.683～0.761 小于基于拉曼光谱的运算结果 0.765～0.830，揭示出二维相关拉曼光谱可以有效地增强实验样品谱峰细节,进一步突出光谱特征的差异。

表 4-4 分别基于拉曼光谱和二维相关拉曼光谱的不同品牌乳制品与
P1 品牌乳制品均值间相关系数运算结果

	P3	P4	P5	P6	P7	P8
P1(拉曼光谱,均值)	0.823	0.830	0.765	0.813	0.825	0.809
P1(二维相关拉曼光谱,均值)	0.747	0.761	0.683	0.740	0.742	0.732

4. 其他品牌乳制品进一步的二维相关拉曼光谱表征及相关系数分析

实验进一步选取 P2 品牌乳制品开展了进一步的研究,图 4-4 显示了采用上述表征方法获得的 P2 品牌乳制品的二维相关拉曼光谱,它大约有如(860,860)等 6 个自动峰和如(860,930)等 30 多个交叉峰,该光谱显示出与前述乳制品有一定差异。采用相关系数分析法对各乳制品之间的相似性进行了定量评价,结果显示:基于拉曼光谱(表 4-5),P2 品牌各乳制品之间的相关系数不小于 0.990;基于二维相关拉曼光谱(表 4-6),P2 品牌各乳制品之间的相关系数不小于 0.985。基于拉曼光谱,P2 品牌各乳制品与其均值之间的相关系数为 0.998±0.001;基于二维相关拉曼光谱,P2 品牌各乳制品与其均值之间的相关系数为 0.997±0.002(表 4-7)。基于拉曼光谱,P3 至 P8 品牌乳制品和 P2 品牌乳制品(均值)间的相关系数位于 0.697~0.962 的范围内;基于二维相关拉曼光谱,P3 至 P8 品牌乳制品和 P2 品牌乳制品(均值)间的相关系数位于 0.596~0.944 的范围内(表 4-8)。

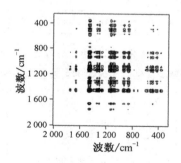

图 4-4 P2 品牌乳制品的同步二维相关拉曼光谱

表 4-5 基于 P2 乳制品各样品间的拉曼光谱相关系数运算结果

	P2-1	P2-2	P2-3	P2-4	P2-5	P2-6
P2-1	1	0.996	0.995	0.997	0.994	0.997
P2-2		1	0.994	0.999	0.990	0.998
P2-3			1	0.995	0.995	0.994
P2-4				1	0.993	0.999
P2-5					1	0.994
P2-6						1

表 4-6 基于 P2 乳制品各样品间的二维相关拉曼光谱相关系数运算结果

	P2-1	P2-2	P2-3	P2-4	P2-5	P2-6
P2-1	1	0.994	0.994	0.995	0.990	0.994
P2-2		1	0.990	0.998	0.985	0.997
P2-3			1	0.991	0.993	0.990
P2-4				1	0.989	0.997
P2-5					1	0.990
P2-6						1

表 4-7 分别基于不同 P2 乳制品的拉曼光谱和二维相关拉曼光谱相关系数运算结果

	P2-1	P2-2	P2-3	P2-4	P2-5	P2-6
P2(拉曼光谱,均值)	0.998	0.999	0.997	0.999	0.996	0.999
P2(二维相关拉曼光谱,均值)	0.997	0.998	0.995	0.998	0.994	0.998

表 4-8 分别基于拉曼光谱和二维相关拉曼光谱的不同品牌乳制品与 P2 品牌乳制品均值间相关系数运算结果

	P1	P3	P4	P5	P6	P7	P8
P2(拉曼光谱,均值)	0.697	0.886	0.937	0.962	0.929	0.861	0.940
P2(二维相关拉曼光谱,均值)	0.596	0.838	0.908	0.944	0.897	0.794	0.912

这些结果表明,P2 品牌各乳制品之间的相似性较高,与其他品牌乳制品的相似性较低。此外,P2 品牌乳制品与 P1 品牌乳制品之间分别基于拉曼光谱和基于二维相关拉曼光谱的相关系数为 0.697 和 0.596。据此,可以获知二维拉曼光谱

可以提高光谱分辨率,能有效地区分各品牌乳制品。

4.1.3 小结

以乳制品鉴别为对象,设计研究了一种新颖的激光微扰二维相关拉曼光谱分析系统,该方案设计较为简便,易于操作,整个实验过程只需几分钟(<5分钟),可以满足快速检测的要求。二维相关拉曼光谱设计,可以增强实验样品的差异性,提高光谱分辨率,提供更加丰富的样品信息,通过相关系数分析,可为定量化区分各品牌乳制品提供工具。因此,该新策略可用于乳制品的质量判别分析,具有潜在的实际应用价值。

4.2 基于高维拉曼光谱的鲜奶制品表征研究

乳制品的质量风险往往来自两个方面:一是非法添加剂或有害物质,二是假冒伪劣产品。传统的识别方法主要包括感官鉴别和仪器分析[1][2][3],但这些方法或存在主观性,或局限于耗时、费力、昂贵的仪器。为了解决现场执法和快速品牌识别问题,近年来以光谱法为代表的快速表征分析方法得到了快速发展。拉曼光谱由于其优越的性能引起了研究人员的广泛关注:便携性强,适用于含水样品的直接测试,光谱信号的采集速度快,可以表征样品丰富的结构信息。然而,研究案例仍主要集中于非法添加剂的检测,如三聚氰胺[4]、非法添加化学药物[5]。现有的表征分析方法只考虑二维拉曼光谱,表征信息较有限。近年来,Noda 教授提出了二

---

[1] Chen X W, Ye N S. Graphene oxide-reinforced hollow fiber solid-phase microextraction coupled with high-performance liquid chromatography for the determination of cephalosporins in milk samples[J]. Food Analytical Methods, 2016, 9(9): 2452-2462.

[2] Gliszczyńska-Świgło A, Chmielewski J. Electronic nose as a tool for monitoring the authenticity of food. A review[J]. Food Analytical Methods, 2017, 10(6): 1800-1816.

[3] Alsammarraie F K, Lin M S. Using standing gold nanorod arrays as surface-enhanced Raman spectroscopy (SERS) substrates for detection of carbaryl residues in fruit juice and milk[J]. Journal of Agricultural and Food Chemistry, 2017, 65(3): 666-674.

[4] Ritota M, Manzi P. Melamine detection in milk and dairy products: Traditional analytical methods and recent developments[J]. Food Analytical Methods, 2018, 11(1): 128-147.

[5] 宁宵,金绍明,李志远,等.功能性老年乳粉中 300 种非法添加药物及其类似物的液相色谱—高分辨质谱分析[J]. 色谱,2023,41(11):960-975.

维相关光谱法[①][②]，这种方法可以提高光谱分辨率，提供更多的光谱细节信息，并呈现出新的分析视角。班晶晶等研究建立了基于表面增强拉曼光谱与二维相关光谱法检测鸡肉中恩诺沙星残留的方法，通过光谱特征筛选与偏最小二乘回归法实现了抗生素残留的检测评估[③]。不过，目前这种方法在乳制品研究领域相关报道还相对较少。在 4.1 节展现了研究小组针对固态乳制品的二维相关光谱的设计与分析，结果呈现出实验样品高维特征信息。本节以鲜奶产品为例，探讨液态乳制品的多维拉曼光谱表征分析，研究基于高维拉曼光谱的乳制品定量鉴别与特征分析方法[④]。

4.2.1 实验部分

1. 材料和拉曼测量

实验用乳制品均购置于南京苏果超市，其中，光明鲜奶产品标记为品牌 P1，卫岗鲜奶产品标记为品牌 P2。

将新鲜牛奶样品（360 μL）置于 96 孔板的单独孔口中，以保持孔口充满。接下来，使用便携式激光拉曼光谱仪记录样品的拉曼光谱，光谱仪型号：Prott-ezRaman-D3，厂家：美国恩威光电公司（Enwave Optronics）。激光的激发波长为 785 nm，激光功率为 450 mW。照射时间从 20 s 依次增加到 180 s，间隔为 40 s。光谱仪在 250～2 000 cm^{-1} 范围内工作，光谱分辨率为 1 cm^{-1}。电荷耦合器件温度：-85 ℃。这些样品在进行拉曼光谱采集时没有进行任何物理和化学预处理。96 孔板：美国康宁公司（Corning Incorporated）。

2. 数据处理

使用拉曼光谱仪自带软件 SLSR Reader V8.3.9 进行拉曼光谱的基线校正。利用 MATLAB 软件（美国 MathWorks 公司）进行小波去噪和欧氏距离计算。二维相关拉曼光谱的计算是使用软件"2D shige"进行［Shigaki Morita，日本的 Kwansei Gakuin University（关西学院大学），2004—2005］。

[①] Noda I, Ozaki Y. Two-Dimensional Correlation Spectroscopy-Applications in Vibrational and Optical Spectroscopy[M]. Chichester：John Wiley& Sons,Inc.，2004：1-195.

[②] Park Y, Jin S L, Noda I, et al. Recent progresses in two-dimensional correlation spectroscopy (2D-COS)[J]. Journal of Molecular Structure, 2018，1168：1-21.

[③] 班晶晶,刘贵珊,何建国,等. 基于表面增强拉曼光谱与二维相关光谱法检测鸡肉中恩诺沙星残留[J]. 食品与机械,2020,36(7)：55-58.

[④] Zhang Z Y, Li S W, Sha M, et al. Characterization of fresh milk products based on multidimensional Raman spectroscopy[J]. Journal of Applied Spectroscopy，2021，87(6)：1206-1215.

4.2.2 结果和讨论

1. 鲜奶制品的拉曼光谱表征分析

鲜奶制品在不同照射时间下的拉曼光谱如图 4-5 所示。参考已有的相关文献报道[1][2][3][4][5],主要光谱振动带及其归属如表 4-9 所示。该谱图提供了样品丰富的分子振动和成分信息,例如,1 762 cm^{-1} 处的拉曼光谱带可以归属于脂肪酸酯的 C=O 伸缩振动,1 668 cm^{-1} 处的谱带可被认为是蛋白质的 C=O 伸缩振动(酰胺 I 的 CONH 基团)和不饱和脂肪酸的 C=C 伸缩振动,1 455 cm^{-1} 处的强拉曼带可能主要是由于脂肪和碳水化合物的 CH_2 变形振动,1 015 cm^{-1} 处是另一个强而显著的拉曼谱带,其可能与蛋白质苯丙氨酸的苯环呼吸振动(环中的 C—C 伸缩振动)有关[6]。与之前报道的奶粉拉曼光谱相比,仅在鲜奶中发现以下拉曼峰:1 212 cm^{-1}、1 195 cm^{-1} 和 1 168 cm^{-1},可能主要与氨基酸有关,这表明奶粉加工后的拉曼特性与鲜奶制品不同[7][8]。图 4-5 显示了 P1 品牌鲜奶样品在不同照射时间下的拉曼光谱变化趋势,可以看出,当测试时间为 20 s 时,只有 1 015 cm^{-1} 和 1 455 cm^{-1} 等少数主峰出现;随着测量时间的延长,出现更多峰并变得清晰,强度依次增加,峰与峰之间的比率基本保持不变。

---

[1] Mendes T O, Junqueira G M A, Porto B L S, et al. Vibrational spectroscopy for milk fat quantification: Line shape analysis of the Raman and infrared spectra[J]. Journal of Raman Spectroscopy, 2016, 47(6): 692-698.

[2] Karacaglar N N Y, Bulat T, Boyaci I H, et al. Raman spectroscopy coupled with chemometric methods for the discrimination of foreign fats and oils in cream and yogurt[J]. Journal of Food and Drug Analysis, 2019, 27(1): 101-110.

[3] Ahmad N, Saleem M. Raman spectroscopy based characterization of desi ghee obtained from buffalo and cow milk[J]. International Dairy Journal, 2019, 89: 119-128.

[4] Amjad A, Ullah R, Khan S, et al. Raman spectroscopy based analysis of milk using random forest classification[J]. Vibrational Spectroscopy, 2018, 99: 124-129.

[5] Mazurek S, Szostak R, Czaja T, et al. Analysis of milk by FT-Raman spectroscopy[J]. Talanta, 2015, 138: 285-289.

[6] Li-Chan E C Y. The applications of Raman spectroscopy in food science[J]. Trends in Food Science & Technology, 1996, 7(11): 361-370.

[7] Almeida M R, de S Oliveira K, Stephani R, et al. Fourier-transform Raman analysis of milk powder: A potential method for rapid quality screening[J]. Journal of Raman Spectroscopy, 2011, 42(7): 1548-1552.

[8] Rodrigues Júnior P H, de Sá Oliveira K, de Almeida C E R, et al. FT-Raman and chemometric tools for rapid determination of quality parameters in milk powder: Classification of samples for the presence of lactose and fraud detection by addition of maltodextrin[J]. Food Chemistry, 2016, 196: 584-588.

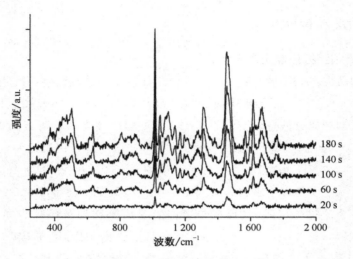

图 4-5 P1 鲜奶制品的拉曼光谱

表 4-9 鲜奶制品拉曼光谱的主要峰值可能归属

波数/cm^{-1}	归属
1 762	$\nu(C=O)_{酯}$
1 668	$\nu(C=O)$酰胺Ⅰ,$\nu(C=C)$
1 596	$\nu(C-C)_{环}$
1 568	$\delta(N-H)$;$\nu(C-N)$酰胺Ⅱ
1 455	$\delta(CH_2)$
1 313	$\tau(CH_2)$
1 275	$\gamma(CH_2)$
1 212	$\nu(C-N)$,酰胺Ⅲ
1 195	$\tau(NH_2)$
1 168	$\nu(C-N)$
1 135	$\nu(C-O)+\nu(C-C)+\delta(C-O-H)$
1 096	$\nu(C-O)+\nu(C-C)+\delta(C-O-H)$
1 045	$\nu(C-O)+\nu(C-C)+\delta(C-O-H)$
1 015	苯环呼吸(苯丙氨酸),$\nu(C-C)_{环}$
964	$\delta(C-O-C)+\delta(C-O-H)+\nu(C-O)$
896	$\delta(C-C-H)+\delta(C-O-C)$
809	$\delta(C-C-O)$

续表 4-9

波数/cm^{-1}	归属
633	$\delta(C—C—O)$
506	葡萄糖
369	乳糖

注：ν—伸缩振动，δ—变形振动，τ—扭曲振动，γ—面外弯曲振动。

2. 基于高维拉曼光谱的鲜奶制品特性分析

实验中采集的鲜奶样品拉曼光谱存在一定的噪声。如图 4-6 所示，噪声是光谱仪器采集信号时产生的随机信号，可能会干扰后续的测量分析。因此，本节选择小波去噪方法对光谱数据进行去噪预处理[①]。简单地说，基本步骤是通过小波对频谱信号进行分解，选择小波基并确定分解级别 N，最后重建平滑的去噪频谱数据。图 4-7 显示了基于 biro2.4 小波基的拉曼光谱分解数据。结果表明，低频信号保留了信号信息，噪声信息主要集中在高频。本节选择了具有三个分解层（$N=3$）的 biro2.4 小波基来实现拉曼光谱数据的去噪。去噪效果如图 4-6 所示，可以清楚地看出，小波去噪后的拉曼光谱信号更平滑。

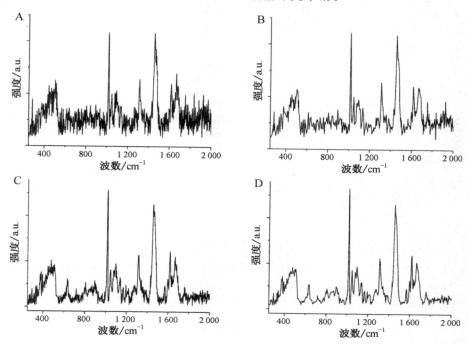

---

① Hoang V D. Wavelet-based spectral analysis[J]. TrAC Trends in Analytical Chemistry, 2014, 62: 144-153.

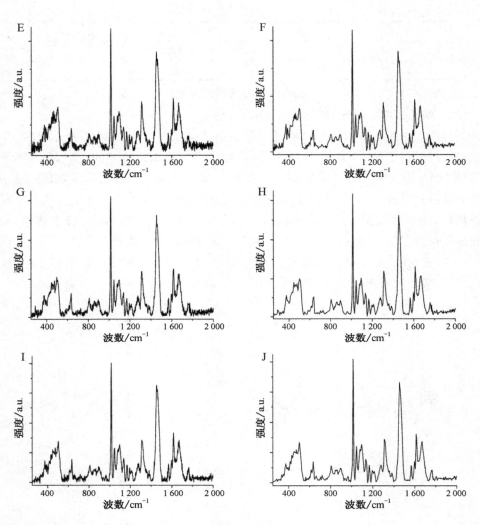

图 4-6 P1 鲜奶制品的照射时间分别为(A)20 s、(C)60 s、(E)100 s、(G)140 s 和 (I)180s 的原始拉曼光谱,以及小波去噪后的照射时间为(B)20 s、(D)60 s、(F)100 s、(H)140 s、(J)180 s 的拉曼光谱

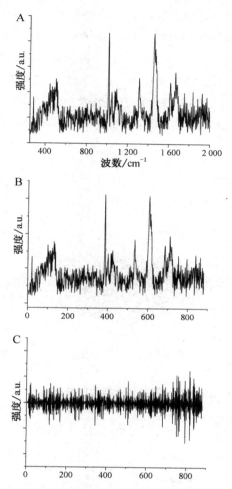

图4-7 P1鲜奶制品照射时间为20 s的原始拉曼光谱(A)，
bior2.4小波去噪分解第一层低频系数(B)，第一层高频系数(C)

二维相关光谱用于构建高维光谱，其可以综合利用多个二维光谱，通过二维相关分析运算获得样品的三维信息。可以有效地提高光谱的分辨率，并提供丰富的样本信息[①]。本节以不同照射时间采集的拉曼光谱信号为外部扰动，构建了样品的二维拉曼相关光谱，三维立体图如图4-8所示，二维平面图如图4-9所示。正相关强度和负相关强度由未出现阴影区域和有阴影区域表示。随着激光照射

---

① Park Y, Noda I, Jung Y M. Novel developments and applications of two-dimensional correlation spectroscopy[J]. Journal of Molecular Structure, 2016, 1124: 11-28.

时间的增加,鲜奶制品的拉曼信号增强,因此相关信号主要是正的。可以看出,在鲜奶样品二维相关光谱信号中,在(1 668,1 668)、(1 612,1 612)、(1 455,1 455)、(1 313,1 313)、(1 096,1 096)、(1 015,1 015)和(506,506)周围有 7 个具有正相关的自动峰,以及在(1 612,1 455)、(1 455,1 313)、(1 455,1 015)、(1 455,1 612)(1 313,1 455)、(1 015,1 455)等具有明显正相关的交叉峰约 60 个。还可以发现,鲜奶样品的两个特征峰(1 015,1 015)和(1 455,1 455)之间存在一定的相关性。

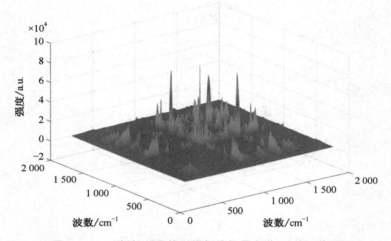

图 4-8　P1 鲜奶制品的二维相关拉曼光谱(三维立体图)

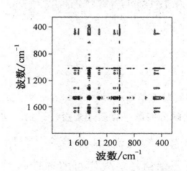

图 4-9　P1 鲜奶制品的二维相关拉曼光谱(二维平面图)

3. 不同品牌鲜奶制品间差异性定量分析

实验组采用 P1 品牌鲜奶,对照组采用 P2 品牌鲜奶,均为乳白色,仅靠人眼无法有效识别。P2 鲜奶制品的拉曼光谱和二维相关拉曼光谱分别如图 4-10(二维拉曼光谱)、图 4-11(三维立体图)和图 4-12(二维平面图)所示。P2 鲜奶制品的拉曼光谱主峰与 P1 鲜奶制品相似,表明 P2 鲜奶制品与 P1 鲜奶的物质组成相似,但同时可以看出拉曼光谱峰的比例不同,表明两个品牌的鲜奶制品在成分含量上存在差异,这也与实际情况一致。如表 4-10 所示,两个品牌的鲜奶制品的主要营养成分在含量上确实存在一定的差异。在二维相关光谱水平上,P2 鲜奶制品的(1 015,1 015)、(1 455,1 455)峰及其比值与 P1 鲜奶制品的峰及其比值显著不同,表明多维拉曼特征峰有望用于识别类似样品。

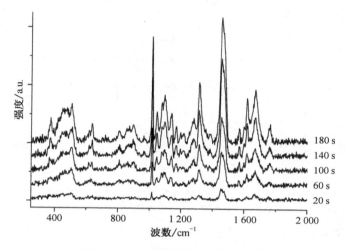

图 4-10 P2 鲜奶制品的拉曼光谱

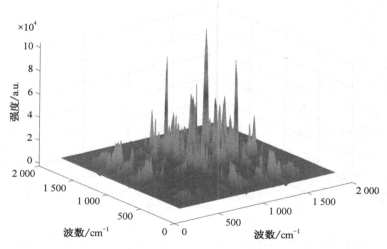

图 4-11 P2 鲜奶制品的二维相关拉曼光谱(三维立体图)

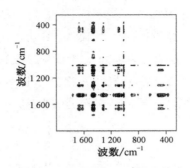

图 4-12　P2 鲜奶制品的二维相关拉曼光谱（二维平面图）

表 4-10　不同品牌鲜奶制品的营养成分和含量信息

营养成分	品牌 P1	品牌 P2
蛋白质	3.0 g/100 g	3.0 g/100 g
脂肪	3.2 g/100 g	3.1 g/100 g
糖类	4.8 g/100 g	4.5 g/100 g

从以上分析可以看出，拉曼光谱和二维相关拉曼光谱可以提供丰富的鲜奶样品成分和分子特征信息，可以从定量分析的角度进一步阐述。欧氏距离计算可用于定量评估样本之间的相似性[1][2]。首先，研究了同一品牌 P1 各样本之间的相似性。随机抽取 6 个鲜奶制品，为了克服鲜奶制品正常质量波动的影响，应用拉曼光谱均值估计其理论值，然后计算各样品与均值之间的欧氏距离，如表 4-11 至表 4-22 所示。在拉曼光谱照射时间为 20 s、60 s、100 s、140 s 和 180 s 的条件下，P1 每个样本与平均值之间的欧氏距离范围分别为 117.938 7～138.941 4（表 4-11）、172.177 9～301.769 7（表 4-12）、222.650 4～423.160 5（表 4-13）、231.641 5～674.057 5（表 4-14）和 335.984 1～596.305 6（表 4-15）。结果表明，同一品牌样品与平均值之间存在一定的质量波动，同时保持较高的一致性。随着拉曼光谱照射时间的变化，在表征样品之间的相似性欧氏距离存在一些波动。基于二维相关拉曼光谱的 P1 品牌鲜奶制品的每个样品与其平均值之间的欧氏距离结果位于 $4.070\,8 \times 10^5 \sim 6.972\,7 \times 10^5$ 的范围内（表 4-16）。结果表明，基

---

[1] He W, Zhou J, Cheng H, et al. Validation of origins of tea samples using partial least squares analysis and Euclidean distance method with near-infrared spectroscopy data[J]. Spectrochimica Acta Part A: Molecular and Biomolecular Spectroscopy, 2012, 86: 399-404.

[2] Chen J B, Zhou Q, Noda I, et al. Discrimination of different Genera Astragalus samples via quantitative symmetry analysis of two-dimensional hetero correlation spectra[J]. Analytica Chimica Acta, 2009, 649(1): 106-110.

于二维相关拉曼光谱的欧氏距离数值较高,这可能是由于二维相关光谱的光谱分辨率较高,维数较大所致。特征提取有望提高判别效率,如前述分析所示,$1\ 015\ cm^{-1}$和$1\ 455\ cm^{-1}$峰值之间的比值可能反映了鲜奶制品的一个主要特征。因此,$1\ 015\ cm^{-1}/1\ 455\ cm^{-1}$的比值被用作进一步评估输入值,以在20 s、60 s、100 s、140 s和180 s的照射时间的拉曼光谱基于拉曼峰比值($1\ 015\ cm^{-1}/1\ 455\ cm^{-1}$处)计算各样本之间的欧氏距离[1][2]。P1品牌鲜奶制品的各样品与其平均值之间的欧氏距离运算结果分别为0.045 6~0.108 5(表4-17)、0.017 1~0.189 5(表4-18)、0.043 3~0.143 7(表4-19)、0.001 1~0.291 7(表4-20)、0.067 4~0.129 3(表4-21)。基于在(($1\ 015,1\ 015$)/($1\ 455,1\ 455$))处的二维相关拉曼光谱峰比值,P1品牌鲜奶制品的各样品与其平均值之间的欧氏距离结果位于0.228 5~0.452 2的范围内(表4-22)。由此可以看出,计算维度覆盖了拉曼光谱(1 751维)、二维相关拉曼光谱(1 751×1 751维),计算结果与之前的结果一致,即样本之间存在一些波动,但保持了较高的一致性。

表4-11 基于拉曼光谱的P1品牌鲜奶制品的每个样本与
其平均值之间的欧氏距离结果(照射时间:20 s)

	P1-1	P1-2	P1-3	P1-4	P1-5	P1-6
P1-20 s(均值)	117.938 7	125.334 5	132.904 5	132.456 9	118.699 3	138.941 4

表4-12 基于拉曼光谱的P1品牌鲜奶制品的每个样本与
其平均值之间的欧氏距离结果(照射时间:60 s)

	P1-1	P1-2	P1-3	P1-4	P1-5	P1-6
P1-60 s(均值)	172.177 9	301.769 7	193.318 6	189.400 8	189.783 8	203.187 1

表4-13 基于拉曼光谱的P1品牌鲜奶制品的每个样本与
其平均值之间的欧氏距离结果(照射时间:100 s)

	P1-1	P1-2	P1-3	P1-4	P1-5	P1-6
P1-100 s(均值)	228.560 7	423.160 5	222.650 4	304.744 3	316.931 4	331.036 2

---

[1] Xie Y, You Q B, Dai P Y, et al. How to achieve auto-identification in Raman analysis by spectral feature extraction & Adaptive Hypergraph[J]. Spectrochimica Acta Part A, Molecular and Biomolecular Spectroscopy, 2019, 222: 117086.

[2] Ahmad N, Saleem M. Characterization of desi ghee obtained from different extraction methods using Raman spectroscopy[J]. Spectrochimica Acta Part A, Molecular and Biomolecular Spectroscopy, 2019, 223: 117311.

表 4-14 基于拉曼光谱的 P1 品牌鲜奶制品的每个样本与
其平均值之间的欧氏距离结果（照射时间：140 s）

	P1-1	P1-2	P1-3	P1-4	P1-5	P1-6
P1-140 s(均值)	231.641 5	523.071 6	291.278 0	424.673 5	470.364 7	674.057 5

表 4-15 基于拉曼光谱的 P1 品牌鲜奶制品的每个样本与
其平均值之间的欧氏距离结果（照射时间：180 s）

	P1-1	P1-2	P1-3	P1-4	P1-5	P1-6
P1-180 s(均值)	492.167 9	382.633 7	335.984 1	596.305 6	369.119 8	328.977 8

表 4-16 基于二维相关拉曼光谱的 P1 品牌鲜奶制品各样本与其平均值之间的欧氏距离结果

	P1-1	P1-2	P1-3	P1-4	P1-5	P1-6
P1-2D(均值)	$6.163\,3\times10^5$	$4.728\,5\times10^5$	$4.070\,8\times10^5$	$6.972\,7\times10^5$	$4.636\,8\times10^5$	$4.444\,9\times10^5$

注：2D 表示二维相关拉曼光谱。

表 4-17 基于拉曼峰($1\,015\,cm^{-1}/1\,455\,cm^{-1}$)比值的拉曼光谱，P1 品牌鲜奶制品的
每个样品与其平均值之间的欧氏距离结果。（照射时间：20 s）

	P1-1	P1-2	P1-3	P1-4	P1-5	P1-6
P1-20 s(均值)	0.054 4	0.062 4	0.106 4	0.045 6	0.062 7	0.108 5

表 4-18 基于拉曼峰($1\,015\,cm^{-1}/1\,455\,cm^{-1}$)比值的拉曼光谱，P1 品牌鲜奶制品的
每个样品与其平均值之间的欧氏距离结果。（照射时间：60 s）

	P1-1	P1-2	P1-3	P1-4	P1-5	P1-6
P1-60 s(均值)	0.087 7	0.189 5	0.062 8	0.097 2	0.017 1	0.076 8

表 4-19 基于拉曼峰($1\,015\,cm^{-1}/1\,455\,cm^{-1}$)比值的拉曼光谱，P1 品牌鲜奶制品的
每个样品与其平均值之间的欧氏距离结果。（照射时间：100 s）

	P1-1	P1-2	P1-3	P1-4	P1-5	P1-6
P1-100 s(均值)	0.062 7	0.143 7	0.070 1	0.104 5	0.107 1	0.043 3

第4章 乳制品高维拉曼光谱数据的构建研究

表 4-20 基于拉曼峰($1\,015\ cm^{-1}/1\,455\ cm^{-1}$)比值的拉曼光谱,P1 品牌鲜奶制品的每个样品与其平均值之间的欧氏距离结果(照射时间:140 s)

	P1-1	P1-2	P1-3	P1-4	P1-5	P1-6
P1-140 s(均值)	0.048 1	0.141 9	0.020 6	0.001 1	0.184 5	0.291 7

表 4-21 基于拉曼峰($1\,015\ cm^{-1}/1\,455\ cm^{-1}$)比值的拉曼光谱,P1 品牌鲜奶制品的每个样品与其平均值之间的欧氏结果(照射时间:180 s)

	P1-1	P1-2	P1-3	P1-4	P1-5	P1-6
P1-180 s(均值)	0.096 6	0.095 2	0.093 4	0.067 4	0.129 3	0.106 9

表 4-22 基于拉曼峰($1\,015,1\,015$)/($1\,455,1\,455$)比值的二维相关拉曼光谱的 P1 品牌鲜奶制品的每个样品与其平均值之间的欧氏距离结果

	P1-1	P1-2	P1-3	P1-4	P1-5	P1-6
P1-2D(均值)	0.262 8	0.269 4	0.335 5	0.228 5	0.452 2	0.377 4

注:2D 表示二维相关拉曼光谱。

随机抽取 P2 品牌鲜奶制品作为对照组,在拉曼光谱、二维相关拉曼光谱和特征提取条件下,计算每个样品与 P1 品牌鲜奶制品平均值之间的欧氏距离,如表 4-23 至表 4-34 所示。基于照射时间 20 s、60 s、100 s、140 s 和 180 s 的拉曼光谱,每个 P2 品牌鲜奶制品与 P1 品牌鲜奶制品平均值之间的欧氏距离结果分别为 147.665 3~221.041 3(表 4-23)、360.916 5~506.250 6(表 4-24)、580.959 5~869.712 4(表 4-25)、857.706 7~$1.174\,5\times10^3$(表 4-26)、902.885 3~$1.664\,5\times10^3$(表 4-27)。基于二维相关拉曼光谱,每个 P2 品牌鲜奶制品与 P1 品牌鲜奶制品平均值之间的欧氏距离结果位于 $1.031\,5\times10^6$~$2.087\,2\times10^6$ 的范围内(表 4-28)。基于照射时间 20 s、60 s、100 s、140 s 和 180 s 的拉曼光谱峰($1\,015\ cm^{-1}/1\,455\ cm^{-1}$)比值,每个 P2 品牌鲜奶制品与 P1 品牌鲜奶制品平均值之间的欧氏距离结果分别为 0.129 3~0.451 1(表 4-29)、0.581 5~0.921 0(表 4-30)、0.393 2~0.713 3(表 4-31)、0.387 4~0.685 9(表 4-32)、0.383 4~0.733 2(表 4-33)。基于($1\,015,1\,015$)/($1\,455,1\,455$)的二维相关拉曼光谱的峰比值,每个 P2 品牌鲜奶制品和 P1 品牌鲜奶制品的平均值之间的欧氏距离位于 1.080 7~1.487 0 的范围内(表 4-34)。上述结果表明,在相同的实验条件下,对照组(P2 品牌鲜奶制品)与实验组(P1 品牌鲜奶制品的平均值)之间的欧氏距离大于实验组(P1 品牌鲜奶制品)中相应样品之间的距离,如图 4-13 所示。结果表明,拉曼

光谱、二维相关拉曼光谱和特征提取与欧氏距离运算相结合,可以定量揭示鲜奶制品各品牌之间的差异情况。

表 4-23 基于拉曼光谱的每个 P2 品牌鲜奶制品与 P1 品牌鲜奶制品平均值之间的欧氏距离结果(照射时间:20 s)

	P2-1	P2-2	P2-3	P2-4	P2-5	P2-6
P1-20 s(均值)	221.041 3	200.373 9	181.903 7	184.941 1	187.932 3	147.665 3

表 4-24 基于拉曼光谱的每个 P2 品牌鲜奶制品与 P1 品牌鲜奶制品平均值之间的欧氏距离结果(照射时间:60 s)

	P2-1	P2-2	P2-3	P2-4	P2-5	P2-6
P1-60 s(均值)	506.250 6	483.245 7	435.916 9	388.133 2	444.318 3	360.916 5

表 4-25 基于拉曼光谱的每个 P2 品牌鲜奶制品与 P1 品牌鲜奶制品平均值之间的欧氏距离结果(照射时间:100 s)

	P2-1	P2-2	P2-3	P2-4	P2-5	P2-6
P1-100 s(均值)	869.712 4	844.308 9	704.463 4	782.879 6	619.277 0	580.959 5

表 4-26 基于拉曼光谱的每个 P2 品牌鲜奶制品与 P1 品牌鲜奶制品平均值之间的欧氏距离结果(照射时间:140 s)

	P2-1	P2-2	P2-3	P2-4	P2-5	P2-6
P1-140 s(均值)	$1.153\ 7 \times 10^3$	$1.164\ 0 \times 10^3$	860.203 0	857.706 7	$1.174\ 5 \times 10^3$	$1.019\ 2 \times 10^3$

表 4-27 基于拉曼光谱的每个 P2 品牌鲜奶制品与 P1 品牌鲜奶制品平均值之间的欧氏距离结果(照射时间:180 s)

	P2-1	P2-2	P2-3	P2-4	P2-5	P2-6
P1-180 s(均值)	$1.533\ 4 \times 10^3$	$1.664\ 5 \times 10^3$	$1.256\ 9 \times 10^3$	$1.329\ 3 \times 10^3$	$1.400\ 7 \times 10^3$	902.885 3

表4-28 基于二维相关拉曼光谱的每个P2品牌鲜奶制品与P1品牌鲜奶制品平均值之间的欧氏距离结果

	P2-1	P2-2	P2-3	P2-4	P2-5	P2-6
P1-2D(均值)	1.5820×10^6	1.7106×10^6	1.3120×10^6	1.5382×10^6	2.0872×10^6	1.0315×10^6

注:2D表示二维相关拉曼光谱。

表4-29 基于拉曼峰($1015\ cm^{-1}/1455\ cm^{-1}$)比值的拉曼光谱,每个P2品牌鲜奶制品和P1品牌鲜奶制品的平均值之间的欧氏距离结果(照射时间:20 s)

	P2-1	P2-2	P2-3	P2-4	P2-5	P2-6
P1-20 s(均值)	0.4312	0.4511	0.3961	0.3734	0.3686	0.1293

表4-30 基于拉曼峰($1015\ cm^{-1}/1455\ cm^{-1}$)比值的拉曼光谱,每个P2品牌鲜奶制品和P1品牌鲜奶制品的平均值之间的欧氏距离结果(照射时间:60 s)

	P2-1	P2-2	P2-3	P2-4	P2-5	P2-6
P1-60 s(均值)	0.5982	0.5815	0.6052	0.7077	0.6671	0.9210

表4-31 基于拉曼峰($1015\ cm^{-1}/1455\ cm^{-1}$)比值的拉曼光谱,每个P2品牌鲜奶制品和P1品牌鲜奶制品的平均值之间的欧氏距离结果(照射时间:100 s)

	P2-1	P2-2	P2-3	P2-4	P2-5	P2-6
P1-100 s(均值)	0.6470	0.7133	0.6102	0.4893	0.5075	0.3932

表4-32 基于拉曼峰($1015\ cm^{-1}/1455\ cm^{-1}$)比值的拉曼光谱,每个P2品牌鲜奶制品和P1品牌鲜奶制品的平均值之间的欧氏距离结果(照射时间:140 s)

	P2-1	P2-2	P2-3	P2-4	P2-5	P2-6
P1-140 s(均值)	0.5948	0.6859	0.5564	0.3874	0.4307	0.5695

表4-33 基于拉曼峰($1015\ cm^{-1}/1455\ cm^{-1}$)比值的拉曼光谱,每个P2品牌鲜奶制品和P1品牌鲜奶制品的平均值之间的欧氏距离结果(照射时间:180 s)

	P2-1	P2-2	P2-3	P2-4	P2-5	P2-6
P1-180 s(均值)	0.7032	0.7332	0.5533	0.5420	0.4840	0.3834

表 4-34 基于拉曼峰(1 015, 1 015)/(1 455, 1 455)比值的二维相关拉曼光谱，每个 P2 品牌鲜奶制品和 P1 品牌鲜奶制品的平均值之间的欧氏距离结果

	P2-1	P2-2	P2-3	P2-4	P2-5	P2-6
P1-2D(均值)	1.423 7	1.487 0	1.207 0	1.143 9	1.080 7	1.229 6

注：2D 表示二维相关拉曼光谱。

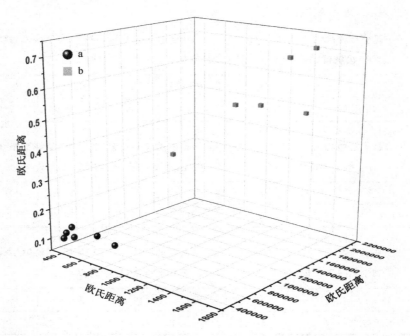

图 4-13 基于欧氏距离的 P1(a)和 P2(b)鲜奶制品的三维分布图
(P1 品牌鲜奶制品的数据，x：表 4-15，y：表 4-16，z：表 4-21，
P2 品牌鲜奶制品数据，x＝表 4-27，y：表 4-28，z：表 4-33)

4.2.3 小结

本节研究工作表明，拉曼光谱、二维相关拉曼光谱和特征提取具有对鲜奶制品进行表征和测量分析的潜力。该方法的优点是多维拉曼光谱可以提供样品成分的丰富分子信息，鲜奶制品作为液体样品无需样品预处理即可直接进行拉曼光谱表征检测，整个检测时间也仅为几分钟。通过调整采样照射时间和二维相关分析，构建了鲜奶制品的二维相关拉曼光谱，从高维水平进一步凸显了样品的光谱特征。基于鲜奶制品的拉曼光谱特征分析，可以通过提取特征峰的比值，实现对鲜奶制品的多层次识别。

第5章　基于相似性度量的乳制品质量判别技术研究

在进行乳制品拉曼光谱数据质量判别分析的过程中,有必要进行样本间的定量化相似性度量。为此,本章选取了拉曼光谱、高维拉曼光谱实验案例,进一步研究展示了相关系数、夹角余弦、欧氏距离等相似性度量方法在乳制品质量判别研究中的重要作用。

5.1　基于拉曼光谱相关系数的乳粉鉴别技术研究

乳粉作为一种常见的大众食品,其质量安全一直受到监管部门和普通消费者的广泛关注[1]。出于经济利益的驱使,假冒伪劣产品时常会混入市场,成为乳粉质量安全的风险因素之一,如偶见于媒体报道的假奶粉事件,给我国乳粉市场的健康发展造成了较为不利的影响[2][3]。目前,乳粉品质分析技术主要体现在两个方面:一方面是我国已建立系列乳制品监管国家标准,如《食品安全国家标准　乳粉》(GB 19644—2010),涵盖了乳粉感官指标、理化指标、微生物限量等指标要求[4],以及《原料乳与乳制品中三聚氰胺检测方法》(GB/T 22388—2008)等检测方法标准[5]。另一方面是新方法、新技术不断涌现,如崔向云等发展的超高效液相色谱-串联质谱法可实现牛乳中舒巴坦的快速测定[6],以及 Zhang 等发展的基于

---

[1] 张淑萍,陆娟.我国乳品行业市场发展整体状况研究[J].中国乳品工业,2013,41(11):33-37.
[2] 刘回春.郴州"大头娃娃"奶粉事件定性:母婴店虚假宣传[J].中国质量万里行,2020(6):68-69.
[3] 史若天.探析公共食品安全事件中政府的舆论引导策略:以2016年上海"假奶粉"事件为例[J].新闻研究导刊,2016,7(12):334.
[4] 中华人民共和国卫生部.食品安全国家标准　乳粉:GB 19644—2010[S].北京:中国标准出版社,2010.
[5] 国家质量监督检验检疫总局.原料乳与乳制品中三聚氰胺检测方法:GB/T 22388—2008[S].北京:中国标准出版社,2008.
[6] 崔向云,常建军,张雪峰,等.UPLC-MS/MS法测定液态乳中舒巴坦残留[J].中国乳品工业,2014,42(10):42-43,45.

表面增强拉曼光谱法可痕量测定牛奶中非法添加物硫氰酸钠等[①]。不过，这些研究方案所关注的主要集中在乳粉的安全性指标评价方面，而在乳粉质量水平评估方面，现有的可行性方案还较为匮乏，难以应对以次充好问题，迫切需要发展具有实际应用价值的乳粉品牌鉴别技术。拉曼光谱结合高通量芯片技术可快速获取奶粉成分散射光谱信息，以全谱作为信号输入，运用相关系数法研究了品牌乳粉相似度差异，提出了具有潜在应用价值的快速鉴别乳粉品牌的方法[②]。

5.1.1 实验

1. 仪器与样品

激光拉曼光谱仪，设备型号：Prott-ezRaman-D3，厂家：美国恩威光电公司（Enwave Optronics），激光波长为 785 nm，激光功率为 450 mW。

实验用乳制品均购置于南京苏果超市，其中，伊利乳粉标记为品牌 P1，雀巢乳粉标记为品牌 P2，美素佳儿乳粉标记为品牌 P3，贝因美乳粉标记为品牌 P4，飞鹤乳粉标记为品牌 P5。

2. 实验过程

使用打孔机制备含有多个圆孔的聚甲基丙烯酸甲酯（polymethyl methacrylate，PMMA）芯片，参数包括：孔直径 8 cm，孔高 9 cm，孔间距 3 cm。在实验过程中，每孔分别上样 100～150 mg 乳粉，震荡使之均匀。而后，使用激光拉曼光谱仪依次采集获取各孔样品的拉曼光谱信息。

3. 相关系数法

相似度计算使用的是相关系数函数，公式如下：

$$相关系数 = \frac{\sum_{i=1}^{n}(x_i - \bar{x}_i)(y_i - \bar{y}_i)}{\sqrt{\sum_{i=1}^{n}(x_i - \bar{x}_i)^2 \sum_{i=1}^{n}(y_i - \bar{y}_i)^2}}$$

式中，x_i 和 y_i 分别表示的是在 i 波长处两样本拉曼光谱强度值，$\bar{x}_i$ 和 $\bar{y}_i$ 分别表示两样本拉曼光谱强度的平均值。

---

① Zhang Z Y, Liu J, Wang H Y. Microchip-based surface enhanced Raman spectroscopy for the determination of sodium thiocyanate in milk[J]. Analytical Letters, 2015, 48(12): 1930-1940.

② 张正勇，沙敏，刘军，等. 基于高通量拉曼光谱的奶粉鉴别技术研究[J]. 中国乳品工业, 2017, 45(6): 49-51.

5.1.2 结果与讨论

1. 乳粉拉曼光谱分析

实验随机选取了市场销售的五种品牌乳粉,分别进行了拉曼光谱信号的采集,如图 5-1 所示。参考已有文献,乳粉拉曼光谱信号可做如下归属分析[1][2][3]:1 748 cm^{-1} 主要源自与脂肪酸有关的酯类 C=O 伸缩振动;1 660 cm^{-1} 主要源自蛋白质酰胺键的 C=O 伸缩振动以及不饱和脂肪酸的 C=C 伸缩振动;拉曼光谱最高峰 1 460 cm^{-1} 主要源自脂肪和糖类相关的 CH_2 变形振动;1 340 cm^{-1} 主要源自 C—O 伸缩振动和 C—H 变形振动;1 308 cm^{-1} 主要源自 CH_2 扭曲振动;1 264 cm^{-1} 主要源自糖类的 CH_2 扭曲振动;1 125 cm^{-1} 和 1 084 cm^{-1} 主要源自糖类相关的 C—O 伸缩振动、C—C 伸缩振动以及 C—O—H 变形振动;1 005 cm^{-1} 主要源自苯丙氨酸的苯环振动;856 cm^{-1} 主要源自 C—C—H 变形振动以及 C—O—C 变形振动,445 cm^{-1} 主要源自 C—C—C 变形振动和 C—O 扭曲振动。由拉曼光谱信号

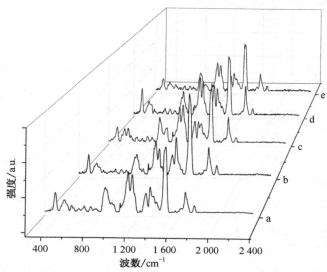

图 5-1 不同品牌 P1(a),P2(b),P3(c),P4(d)和 P5(e)乳粉的拉曼光谱图

---

[1] Almeida M R, de S Oliveira K, Stephani R, et al. Fourier-transform Raman analysis of milk powder: A potential method for rapid quality screening[J]. Journal of Raman Spectroscopy, 2011, 42(7): 1548-1552.

[2] Moros J, Garrigues S, de la Guardia M. Evaluation of nutritional parameters in infant formulas and powdered milk by Raman spectroscopy[J]. Analytica Chimica Acta, 2007, 593(1): 30-38.

[3] Mazurek S, Szostak R, Czaja T, et al. Analysis of milk by FT-Raman spectroscopy[J]. Talanta, 2015, 138: 285-289.

分析可知,各品牌乳粉的拉曼光谱信号主要归结于蛋白质、脂肪和糖类等乳粉主要成分,因而在拉曼光谱信号总体表现情况上出峰位置较为相似,但由于生产工艺和加工原料的不同,各品牌乳粉的拉曼光谱信号在峰型、峰强度上又表现出一定差异,借助相关系数法[①],有望实现各品牌乳粉拉曼光谱差异的量化表征。

2. P1品牌乳粉相关系数分析

进一步,研究以P1品牌乳粉为对象,考察品牌乳粉内部相似度差异情况。随机选取了10个P1品牌乳粉样本,利用自制高通量多孔芯片,采集获得了各样本的拉曼光谱信息。由于产品质量存在一定波动,其表征信号的数学期望为样本信号均值,因此,可用样本测试信号的均值作为P1品牌乳粉的特征信号,而各样本实测信号将围绕均值信号上下波动。基于这一假定,研究比较了各测试样本的拉曼光谱信号与均值信号的相似度变化情况,如图5-2所示。结果显示出,各P1品牌乳粉样本的拉曼光谱信号与P1品牌乳粉均值信号间表现出相关系数0.99以上的高度相似现象。这可能是由于同一品牌各测试样本间的加工工艺、原料等保持了较高的一致性,因而各样本与均值信号间相似程度较高。基于这一现象,进一步运用质量波动控制图方法,尝试建立品牌乳粉的质量控制图,如图5-3所示。质量波动控制图是一种统计过程控制的数理分析手段,基于统计假设检验原理构造而成。当质量波动仅有偶然因素影响时,质量波动处于统计控制状态,即处于上下控制限内围绕中心线波动;而当质量波动受到异常因素影响时,质量波动处于失控状态,质量波动将跃出控制限,故其可起到及时预警作用。质量波动控制图早期主要在机械制造领域得到广泛使用,目前,在食品控制领域也正日益发挥出重要作用[②③]。依据控制图绘制原理,得到了P1品牌乳粉相关系数值变化控制图以及移动极差控制图,由图5-3A可以看出,测试样本的相关系数值围绕中心线(均值)0.9995,在上控制限0.9997和下控制限0.9992间波动。需要补充一点,与一般用控制图不同,在本实验体系考察的是相关系数值,若出现相关系数大于0.9997即超出上控制限的情况时,并不能据此认为出现异常因素,因为相关系数值愈接近于1,说明产品质量的测试信号愈接近于均值,是质量水平提升的表现。此外,图5-3B中相应移动极差值也在理论范围内波动。

---

① Chen J B, Zhou Q, Noda I, et al. Quantitative classification of two-dimensional correlation spectra[J]. Applied Spectroscopy, 2009, 63(8): 920-925.
② 王海燕,张庆民.质量分析与质量控制[M].北京:电子工业出版社,2015:1-175.
③ 刘锐,魏益民,张波.基于统计过程控制(SPC)的挂面加工过程质量控制[J].食品科学,2013, 34(8):43-47.

第 5 章 基于相似性度量的乳制品质量判别技术研究

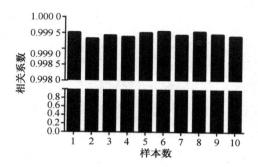

图 5-2 P1 品牌乳粉各样本与其均值间的拉曼光谱相关系数值

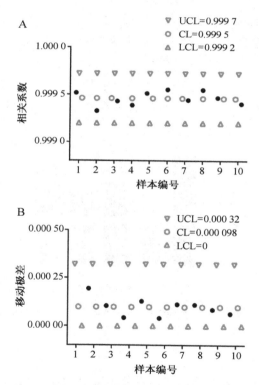

图 5-3 P1 品牌乳粉相关系数值变化控制图(A),P1 品牌乳粉相关系数值移动极差控制图(B)
　　注:UCL 表示上控制限,LCL 表示下控制限,CL 表示中心线。

3. 乳粉相关系数分析

研究进一步考察了不同品牌乳粉间的拉曼光谱相似度情况,以 P1 品牌乳粉(均值)为对象,比较了其他品牌乳粉与之相关系数值差异,如图 5-4 所示。P2、P3、P4、P5 品牌乳粉与 P1 品牌乳粉的拉曼光谱相关系数值分别为 0.950、0.977、0.971、0.995,可以看出各品牌奶粉与研究对象 P1 品牌乳粉间存在较高的拉曼光谱相似度,这可归结于与乳粉主要成分均为蛋白质、脂肪、糖类有关。但是,各品牌乳粉相似度值也存在一定差异,这可能归结于其生产工艺、原料差异导致最终产品间的微量成分不同。进一步分析,P2、P3 与 P1 品牌乳粉相关系数值差异较大,可能原因在于这两种产品为海外品牌,P1 品牌乳粉为国内品牌,原料产地差异较大,存在一定影响。P4、P5 均为国内品牌,P4 与 P1 品牌乳粉相关系数值小于 P5 品牌乳粉,原因可能是在奶源产地上,与 P4、P5 品牌乳粉分别与 P1 品牌乳粉的奶源地理纬度差异不同有关。尽管各品牌乳粉与 P1 品牌乳粉间的相关系数值达到了 0.95 以上,不过与 P1 品牌乳粉内部拉曼光谱相关系数值比较而言,这种不同品牌乳粉的相关系数值差异依旧是明显小于图 5-3A 的下控制限 0.999 2,因此,实验方法依然能够有效地鉴别出品牌乳粉的相似度差异,揭示出实验方法在乳粉品牌间具有一定的区分应用价值。

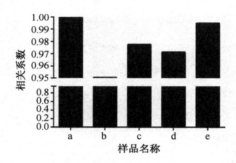

图 5-4 不同品牌 P2(b)、P3(c)、P4(d)和 P5(e)乳粉与 P1(a)乳粉的拉曼光谱相关系数值

5.1.3 小结

本节针对品牌乳粉在产销过程中的相似样品判别需求,以多孔芯片为平台,高通量获取乳粉拉曼光谱信息,结合相关系数运算,量化输出了品牌乳粉内部及品牌乳粉间相似度差异情况,结果显示品牌乳粉内部相关系数值差异较小,品牌乳粉间相关系数值差异相对较大,并通过建立乳粉相关系数质量波动控制图,实现了品牌乳粉动态监控,结合相关系数运算,可初步实现快速、高效的乳粉品牌鉴别,揭示了方法的潜在应用价值。

5.2 基于拉曼光谱相似性度量的乳粉高维表征及判别分析

乳制品作为消费者日常食品购买的重要组成部分,尤其是满足婴幼儿营养需求的重要食品来源,其质量安全问题一直深受我国市场主管部门和大众的关注,然而近年来,国内外乳制品相关质量安全问题仍时见报道,为此,《中华人民共和国国民经济和社会发展第十四个五年规划和 2035 年远景目标纲要》明确提出要加强和改进食品安全监管制度,完善食品安全法律法规和标准体系,深入实施食品安全战略,着力提高食品安全检测技术[1]。针对乳制品质量安全风险,目前管控和研究策略主要包括:(1) 建立健全国家乳制品相关产品标准、检测标准;(2) 研究开发各类先进检测方法。如李涛等研究建立了一种基于超高效液相色谱-质谱的乳粉中 16 种喹诺酮类药物的检测方法[2],de Oliveira Mendes 等运用拉曼光谱结合主成分分析、偏最小二乘回归法进行了生牛奶中乳清的检测[3]。Li 等研究构建了一种基于表面增强拉曼光谱的侧向流免疫传感器,可实现牛乳中粘菌素的快速识别[4]。

不过现有研究主要集中于乳制品相关具体指标成分的测定方面,在乳制品整体品质控制方面的研究案例还相对较为匮乏,突出表现在目前"海淘"过程中伪品、次品时有出现,以次充好"假乳粉"事件等,而伪次品各项指标可能符合国家最低限定标准,造成了新的监管困境。因此,迫切需要发展乳制品品质快速判别的新方法。本节提出了乳制品特征谱图分析思路,以牛、羊乳粉为例,从传统拉曼光谱、二维相关拉曼光谱两个方面构建牛、羊乳粉特征谱图,并进一步结合相似度分析算法量化分析了牛、羊乳粉谱图间的差异[5]。

---

[1] 王立甜. 聚焦"十四五"食品安全[N]. 中国市场监管报,2021 - 12 - 07(3).

[2] 李涛,周艳华,向俊,等. 分散固相萃取—超高效液相色谱—三重四极杆质谱法快速检测奶粉中 16 种喹诺酮药物残留[J]. 中国乳品工业,2021,49(11):54 - 58,64.

[3] de Oliveira Mendes T, Manzolli Rodrigues B V, Simas Porto B L, et al. Raman Spectroscopy as a fast tool for whey quantification in raw milk[J]. Vibrational Spectroscopy, 2020, 111: 103150.

[4] Li Y, Tang S S, Zhang W J, et al. A surface-enhanced Raman scattering-based lateral flow immunosensor for colistin in raw milk[J]. Sensors and Actuators B: Chemical, 2019, 282: 703 - 711.

[5] 顾颖,赵姝彤,熊蓝萍,等. 基于多维拉曼光谱的乳粉表征及判别分析[J]. 粮食科技与经济,2022,47(3):97 - 101.

5.2.1 实验部分

1. 材料

实验用牛、羊乳粉均购置于南京苏果超市。其中,牛乳粉标记为品牌 P1,羊乳粉标记为品牌 P2。

2. 仪器与设备

激光拉曼光谱仪,光谱仪型号:Prott-ezRaman-D3,厂家:美国恩威光电公司(Enwave Optronics),激光波长为 785 nm,激光最大功率约为 450 mW,电荷耦合器件检测器温度控制在 -85 ℃,照射时间为 50 s,扫描次数为 1 次,光谱采集范围为 $250\sim 2\,000$ cm^{-1},光谱分辨率为 1 cm^{-1}。96 孔板:美国康宁公司(Corning Incorporated)。

3. 谱图数据采集方法

取适量乳粉粉末样品置于 96 孔板的各自独立小孔内,使得小孔恰好处于充满状态。而后,使用激光拉曼光谱仪光纤探头对准样品直接进行照射测试,收集信号,即得到样品的拉曼光谱数据。调节激光功率,分别在约 255 mW、320 mW、385 mW、450 mW 条件下获取实验样品的拉曼光谱数据,用以构建二维相关拉曼光谱图。

4. 谱图数据分析方法

拉曼光谱数据经光谱仪自带软件进行基线校正(SLSR Reader V8.3.9,Enwave Optronics,美国恩威光电公司)、小波降噪处理(MATLAB 软件,美国 MathWorks 公司),而后,使用 2D-Shige 软件(Shigeaki Morita, Kwansei Gakuin University,日本关西学院大学)进行动态拉曼光谱数据的二维相关分析,得到样品的二维相关拉曼光谱图。

针对牛、羊乳粉的拉曼光谱和二维相关拉曼光谱数据,分别从相似性相似度、差异性相似度两个方面进行样品的相似度分析。相似性相似度是衡量各光谱内在化学成分配比的相似程度,研究选取了相关系数法、向量夹角余弦法;差异性相似度是衡量两个谱图间的差异,研究选取了欧氏距离法。运算平台:MATLAB R2018a(美国 MathWorks 公司)。

5.2.2 结果与分析

1. 牛、羊乳粉的拉曼光谱分析

牛、羊乳粉均是市场上较为常见的乳制品,不过品种间也存在着一定的差异,有研究显示,羊乳不含牛乳中某些可能导致过敏的异性蛋白,脂肪和蛋白质颗粒也相对较小以至更易于被人体消化吸收,羊乳具有乳蛋白、免疫球蛋白含量较高,脂肪碳链较短等特点,受到消费者的广泛认可[1]。不过羊乳的产量较牛乳要少很多,因而目前乳制品市场主体仍是以牛乳制品为主,在价格上,普遍表现出羊乳价格高于牛乳价格的现象,因此,有必要建立牛、羊乳制品的特征谱图,加强对牛、羊乳的品质管控。拉曼光谱是近年来得到飞速发展的一种散射光谱技术,具有谱图数据采集速度快、样品前处理耗时少甚至无需样品前处理等特点,并且拉曼光谱仪方便携常,十分适用于现场快速检测[2]。实验以牛、羊乳粉为研究对象,采集了其在不同激光功率条件下的拉曼光谱信号,如图5-5和图5-6所示。首先,可以参考已有相关文献报道对谱图中的光谱峰进行物质归属分析[3][4][5],如在1 750 cm^{-1}出现的拉曼光谱峰主要可能是源于与脂肪有关的酯C=O伸缩振动,牛乳的光谱最高峰出现在1 460 cm^{-1},其主要是与源自糖类、脂肪的CH_2变形振动有关,而羊乳的光谱最高峰出现在1 090 cm^{-1},主要是与源自糖类有关的C—O伸缩振动、C—C伸缩振动以及C—O—H变形振动,在1 010 cm^{-1}处出现的拉曼光谱峰是一个较为特殊的峰,其主要是与样品所含的蛋白质中氨基酸苯丙氨酸的苯环振动有关。此外,通过谱图分析还可发现,羊乳在1 025 cm^{-1}处出现的光谱峰较牛乳更为尖锐,羊乳在882 cm^{-1}和385 cm^{-1}处有出现光谱峰,而牛乳在此两处没有明显的光谱峰,更多的光谱峰归属可见表5-1。由此可以看出,通过拉曼光谱表征可以显示出牛、羊乳粉丰富的物质组成信息,主要含有脂肪、糖类和蛋白质分子,有个别拉曼光谱峰仅羊乳显现,在拉曼光谱的峰型、峰位置、峰强度方面,牛、羊乳粉谱图表现出明显不同,反映出牛、羊乳物质组成有一定差异。其次,从图中还可以看出,在不同激光功率条件下,牛、羊乳样品的拉曼光谱峰呈现出一定的变

---

[1] dos Santos Pereira E V, de Sousa Fernandes D D, de Araújo M C U, et al. Simultaneous determination of goat milk adulteration with cow milk and their fat and protein contents using NIR spectroscopy and PLS algorithms[J]. LWT, 2020, 127: 109427.

[2] 张正勇,李丽萍,岳彤彤,等.乳粉拉曼光谱表征数据的标准化与降噪处理研究[J].粮食科技与经济,2018,43(6):57-61.

[3] 王梓笛,李双妹,尹延东,等.基于支持向量机算法的乳制品分类识别技术研究[J].粮食科技与经济,2020,45(3):104-107.

[4] Genis D O, Sezer B, Durna S, et al. Determination of milk fat authenticity in ultra-filtered white cheese by using Raman spectroscopy with multivariate data analysis[J]. Food Chemistry, 2021, 336: 127699.

[5] 罗洁,王紫薇,宋君红,等.不同品种牛乳脂质的共聚焦拉曼光谱指纹图谱[J].光谱学与光谱分析,2016,36(1):125-129.

化规律,比如在约 255 mW 时,仅有少量的谱峰出现;随着激光功率增大,大量的谱峰开始出现,且谱峰强度明显增强,表现出一定的动态变化特征。

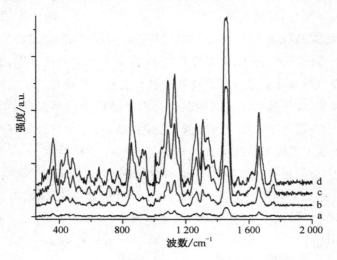

图 5-5 不同激光功率下牛乳粉的拉曼光谱图
(a:～255 mW;b:～320 mW;c:～385 mW;d:～450 mW)

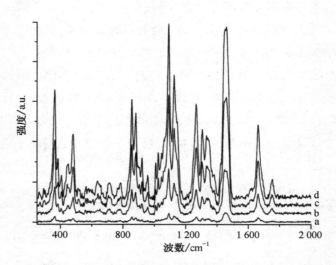

图 5-6 不同激光功率下羊乳粉的拉曼光谱图
(a:～255 mW;b:～320 mW;c:～385 mW;d:～450 mW)

表 5-1　牛、羊乳粉主要的拉曼光谱峰可能归属

牛乳拉曼光谱峰 （波数/cm^{-1}）	羊乳拉曼光谱峰 （波数/cm^{-1}）	归属
1 752	1 754	$\nu(C=O)_{酯}$
1 663	1 663	$\nu(C=O)$(酰胺Ⅰ),$\nu(C=C)$
1 460	1 460	$\delta(CH_2)$
1 379	1 383	$\delta(C-H);\nu(C-O)$
1 344	1 335	$\delta(C-H);\nu(C-O)$
1 309	1 307	$\tau(CH_2)$
1 266	1 268	$\gamma(CH_2)$
1 130	1 127	$\nu(C-O)+\nu(C-C)+\delta(C-O-H)$
1 086	1 091	$\nu(C-O)+\nu(C-C)+\delta(C-O-H)$
1 019	1 025	$\nu(C-O)+\nu(C-C)+\delta(C-O-H)$
1 008	1 009	$\nu(C-C)_{苯环振动(苯丙氨酸)}$
945	958	$\delta(C-O-C)+\delta(C-O-H)+\nu(C-O)$
927	923	$\delta(C-O-C)+\delta(C-O-H)+\nu(C-O)$
	882	$\delta(C-C-H)+\delta(C-O-C)$
856	856	$\delta(C-C-H)+\delta(C-O-C)$
768	783	$\delta(C-C-O)$
719	713	$\nu(C-S)$
650	637	$\delta(C-C-O)$
484	480	$\delta(C-C-C)+\tau(C-O)$
452	450	$\delta(C-C-C)+\tau(C-O)$
413	408	葡萄糖
	385	乳糖
359	364	乳糖

注：ν 为伸缩振动；δ 为变形振动；τ 为扭曲振动，γ 为面外弯曲振动。

2. 牛、羊乳粉的二维相关拉曼光谱分析

前述分析揭示出牛、羊乳粉拉曼光谱随着外界条件变化呈现出一定的动态变化特征，据此可进一步开展二维相关分析，该技术是由 Isao Noda 教授等提出的一种新型光谱数据处理方案，基本思路就是收集样品在外扰作用下动态谱图，而

后运用相关分析方法,实现多张谱图的重构融合,构建样品在三维空间中的特征谱图[1][2]。为减少光谱采集过程中随机噪声对后续三维谱图构建的影响,实验采用小波降噪的方法进行了光谱预处理,选择coif1小波基,有效消除了光谱噪声。进一步对上述降噪后的四张光谱的数据进行处理,以平均峰为参考峰,运用二维相关分析法进行三维谱图构建,如图5-7所示,为牛乳粉的二维相关拉曼光谱同步图,显示其最大的自动峰为(1 666,1 666),其他自动峰有(1 455,1 455)、(1 326,1 326)、(1 100,1 100)、(856,856)等,交叉峰有(1 455,1 660)、(1 290,1 440)、(1 100,1 290)、(870,1 110)等。由于动态光谱信号是随着激光强度增强依次增大,故在新构建的三维光谱图中相关光谱信号均为正相关,数值为正值。采用同样的处理方法,得到羊乳粉的二维相关拉曼光谱同步图,如图5-8所示,自动峰有(1 666,1 666)、(1 455,1 455)、(1 265,1 265)、(1 100,1 100)、(870,870)、(366,366)等,交叉峰有(1 455,1 666)、(1 300,1 445)、(1 100,1 300)、(1 100,1 460)、(876,1 090)、(870,1 455)等,可以明显看出,牛、羊乳粉在三维空间特征谱图上差异显著,羊乳粉的出峰数量明显较多。

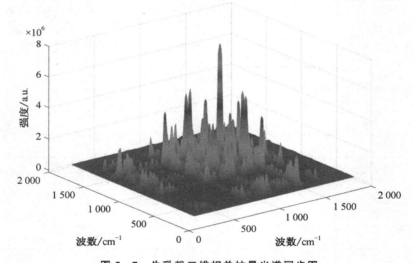

图5-7 牛乳粉二维相关拉曼光谱同步图

---

[1] Noda I, Ozaki Y. Two-Dimensional Correlation Spectroscopy-Applications in Vibrational and Optical Spectroscopy[M]. Chichester: John Wiley & Sons, Inc., 2004: 1-195.

[2] Marcott C, Kansiz M, Dillon E, et al. Two-dimensional correlation analysis of highly spatially resolved simultaneous IR and Raman spectral imaging of bioplastics composite using optical photothermal Infrared and Raman spectroscopy[J]. Journal of Molecular Structure, 2020, 1210: 128045.

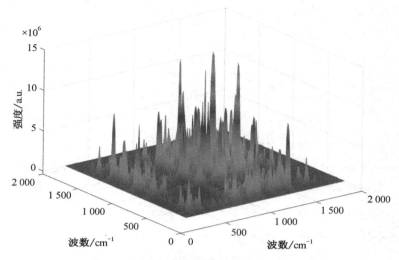

图 5-8 羊乳粉二维相关拉曼光谱同步图

3. 牛、羊乳粉的拉曼光谱相似度分析

基于拉曼光谱、二维相关拉曼光谱可表征样品丰富的化学组成、光谱特征信息,并可进行直观的人眼鉴别,但是尚难以给出样品间定量化的判别分析,因此,实验进一步结合相似度算法,进行量化分析。实验随机选取了 3 个牛乳粉样品数据和 3 个羊乳粉样品数据,分别计算该 6 个样品数据之间的相似度。以各样品在 450 mW 条件下测试获取拉曼光谱并经小波降噪后的数据为输入,用相关系数法计算得到牛乳粉样品间相关系数值为 0.996 5~0.998 2,羊乳粉样品间相关系数值为 0.985 7~0.998 6,而牛、羊乳粉样品间相关系数值为 0.947 0~0.972 2(表 5-2)。向量夹角余弦系数计算得到牛乳粉样品间夹角余弦值为 0.997 7~0.998 8,羊乳粉样品间夹角余弦值为 0.990 8~0.999 1,牛、羊乳粉样品间夹角余弦值为 0.965 8~0.981 7(表 5-3)。欧氏距离计算得到牛乳粉样品间欧氏距离为 0.428 6~0.881 3($\times 10^4$),羊乳粉样品间欧氏距离为 1.028 9~1.434 1($\times 10^4$),羊、牛乳粉样品间欧氏距离为 2.080 9~3.847 9($\times 10^4$)(表 5-4)。

以样品二维相关拉曼光谱数据为输入,相关系数法计算得到牛乳粉样品间相关系数值为 0.994 9~0.997 3,羊乳粉样品间相关系数值为 0.978 2~0.997 9,而牛、羊乳粉样品间相关系数值为 0.921 9~0.958 0(表 5-5)。向量夹角余弦系数计算得到牛乳粉样品间夹角余弦值为 0.995 5~0.997 6,羊乳粉样品间夹角余弦值为 0.980 5~0.998 1,牛、羊乳粉样品间夹角余弦值为 0.931 0~0.962 6(表 5-6)。欧氏距离计算得到牛乳粉样品间欧氏距离为 0.639 5~1.632 5($\times 10^8$),

羊乳粉样品间欧氏距离为 2.019 8～3.097 2(×10^8),牛、羊乳粉样品间欧氏距离为 4.679 0～8.510 3 (×10^8)(表 5-7)。

表 5-2　牛羊乳拉曼光谱相关系数计算结果

	牛乳1	牛乳2	牛乳3	羊乳1	羊乳2	羊乳3
牛乳1	1.000 0	0.996 5	0.997 9	0.972 2	0.947 9	0.953 1
牛乳2		1.000 0	0.998 2	0.970 9	0.950 3	0.955 5
牛乳3			1.000 0	0.971 7	0.947 0	0.952 5
羊乳1				1.000 0	0.985 7	0.987 9
羊乳2					1.000 0	0.998 6
羊乳3						1.000 0

表 5-3　牛羊乳拉曼光谱夹角余弦计算结果

	牛乳1	牛乳2	牛乳3	羊乳1	羊乳2	羊乳3
牛乳1	1.000 0	0.997 7	0.998 6	0.981 7	0.966 3	0.969 5
牛乳2		1.000 0	0.998 8	0.980 9	0.968 0	0.971 2
牛乳3			1.000 0	0.981 4	0.965 8	0.969 2
羊乳1				1.000 0	0.990 0	0.992 0
羊乳2					1.000 0	0.999 1
羊乳3						1.000 0

表 5-4　牛羊乳拉曼光谱欧氏距离计算结果(10^4)

	牛乳1	牛乳2	牛乳3	羊乳1	羊乳2	羊乳3
牛乳1	0.000 0	0.428 6	0.714 6	2.198 7	2.409 5	3.310 7
牛乳2		0.000 0	0.881 3	2.080 9	2.271 1	3.144 7
牛乳3			0.000 0	2.737 6	2.867 4	3.847 9
羊乳1				0.000 0	1.028 9	1.434 1
羊乳2					0.000 0	1.157 9
羊乳3						0.000 0

表5-5　牛羊乳二维相关拉曼光谱相关系数计算结果

	牛乳1	牛乳2	牛乳3	羊乳1	羊乳2	羊乳3
牛乳1	1.000 0	0.994 9	0.997 0	0.958 0	0.922 9	0.930 5
牛乳2		1.000 0	0.997 3	0.955 9	0.926 4	0.933 6
牛乳3			1.000 0	0.957 4	0.921 9	0.929 8
羊乳1				1.000 0	0.978 2	0.981 7
羊乳2					1.000 0	0.997 9
羊乳3						1.000 0

表5-6　牛羊乳二维相关拉曼光谱夹角余弦计算结果

	牛乳1	牛乳2	牛乳3	羊乳1	羊乳2	羊乳3
牛乳1	1.000 0	0.995 5	0.997 3	0.962 6	0.931 9	0.938 5
牛乳2		1.000 0	0.997 6	0.960 7	0.935 0	0.941 3
牛乳3			1.000 0	0.962 0	0.931 0	0.937 9
羊乳1				1.000 0	0.980 5	0.983 7
羊乳2					1.000 0	0.998 1
羊乳3						1.000 0

表5-7　牛羊乳二维相关拉曼光谱欧氏距离计算结果（10^8）

	牛乳1	牛乳2	牛乳3	羊乳1	羊乳2	羊乳3
牛乳1	0.000 0	0.639 5	1.337 6	4.945 5	4.925 1	7.429 0
牛乳2		0.000 0	1.632 5	4.719 2	4.679 0	7.151 9
牛乳3			0.000 0	6.057 9	5.894 4	8.510 3
羊乳1				0.000 0	2.019 8	3.097 2
羊乳2					0.000 0	2.818 4
羊乳3						0.000 0

定量化的分析结果揭示出：(1) 无论是相关系数、夹角余弦，还是欧氏距离，结果均显示出牛、羊乳粉同品种样品间具有较高的相似度，而牛、羊乳粉不同品种样品间则表现出较大的差异度。同时，较之相关系数、夹角余弦值，欧氏距离在刻画牛、羊乳粉样品间的差异性上更为明显。(2) 与直接使用拉曼光谱结合相似度运算相比，二维相关拉曼光谱在揭示牛、羊乳粉样品内相似、样品间差异方面，可提供更高的谱图分辨率，因而相关系数、夹角余弦值表现出下降的趋势，欧氏距离

值则表现出增大的趋势。

5.2.3 小结

实验提出并构建了牛、羊乳粉的多维拉曼光谱特征谱图表征方案,系统论证并定量化评估了牛、羊乳粉同品种样品间相似、不同品种样品间差异的情况。拉曼光谱呈现了样品丰富的特征组分信息,二维相关拉曼光谱则进一步呈现了样品动态变化特征,进一步结合相关系数、夹角余弦和欧氏距离算法,定量化分析结果揭示出牛、羊乳粉间的差异性较大,显示出二维相关拉曼光谱图分辨率更高,可实现样品有效判别。该实验方案具备在牛、羊乳粉鉴别、质量控制等多个方面的潜在应用价值,并且这种多维特征谱图构建和多层次分析思路也可为其他食品的质量安全监管提供借鉴。

第 6 章　基于智能判别算法的乳制品质量智能判别技术研究

通过前述研究可以获知拉曼光谱可以表征乳制品丰富的组分信息,而智能学习算法具备运算速度快、量化客观评价结果等特点,两者的有效结合成为质量判别领域重要的研究方向。本章以实验案例的形式进一步研究展示了 k 近邻算法、支持向量机算法、随机森林算法、概率神经网络算法、极限学习机算法与乳制品拉曼光谱数据相结合的判别应用情况,并进行了参数优化研究。

6.1　基于拉曼光谱与 k 近邻算法的酸奶鉴别研究

酸奶是人们生活中常见的乳制品之一,常常是以牛奶为原料,经杀菌处理后,向牛奶中添加适量的有益菌、发酵剂,经过一段时间的发酵后,再冷却灌装的一种乳制产品。酸奶营养丰富,易于吸收,深受消费者喜爱。近年来,食品质量安全问题时有发生,如偶见媒体报道的以次充好假冒品牌奶粉事件、3·15 曝光的各类食品质量安全风险事件等,使得食品质量管理持续成为社会关注的高频热点问题之一。《食品安全国家标准　发酵乳》(GB 19302—2010)从酸奶的感官要求、理化指标、微生物限量、乳酸菌数等进行了详细的指标规定,不过这些指标的检测需要一定的时间和较为繁琐的实验操作[1]。拉曼光谱作为一种新型光谱表征技术,可以获得测试对象分子振动信息,具有测试速度快、谱峰信息丰富等特点[2]。与近红外光谱相比较而言,拉曼光谱的谱峰可以予以准确归属,分子振动信息物质基础明确,与红外光谱易受到水的干扰不同,水的拉曼散射截面小,对于拉曼光谱采集没有明显干扰,而酸奶一般表现为黏稠状液体,因此,拉曼光谱表征酸奶时,

---

[1]　中华人民共和国卫生部.食品安全国家标准　发酵乳:GB 19302—2010[S].北京:中国标准出版社,2010.

[2]　Hua M Z, Feng S L, Wang S, et al. Rapid detection and quantification of 2, 4-dichlorophenoxyacetic acid in milk using molecularly imprinted polymers-surface-enhanced Raman spectroscopy [J]. Food Chemistry, 2018, 258: 254-259.

可以直接上样采集,避免了制样前处理,操作简便快捷,成为酸奶质量控制的优越备选技术之一。此外,随着机器学习算法的普及与发展,拉曼光谱快速采集技术结合机器学习辅助智能判别分析技术成为酸奶质量控制新体系的重要组成部分。本节以三种市售品牌酸奶的鉴别分析为例,采集与分析酸奶的拉曼光谱,结果显示品牌酸奶间具有较高的相似性,仅通过人工解谱已经很难予以有效区分,于是,发展与优化了 k 近邻算法,通过智能识别算法可以快速、高效地予以判别分析。每个样品的拉曼光谱采集时间仅为 100 s,数据分析时间仅需 1 s,优化模型识别率达到 99% 以上,因此,论证了一种可望用于酸奶质量控制的快速鉴别方法[1][2]。

6.1.1 实验部分

1. 材料

实验用酸奶均购置于南京苏果超市,三个品牌酸奶均为原味口味,其中,光明酸奶标记为品牌 P1,蒙牛酸奶标记为品牌 P2,伊利酸奶标记为品牌 P3,每个品牌随机采样 20 个,样品共计 60 个。

2. 仪器与设备

拉曼光谱采集使用的是便携式激光拉曼光谱仪,光谱仪型号:Prottez Raman-D3,厂家:美国恩威光电公司(Enwave Optronics),仪器激光波长为 785 nm,激光最大功率约为 450 mW,光谱范围为 250~2 339 cm^{-1},光谱分辨率为 1 cm^{-1},光谱采集时间 100 s。实验用酸奶上样装置为 96 孔板,厂家:美国康宁公司(Corning Incorporated)。

3. 拉曼光谱采集

取适量酸奶样品置于 96 孔板的独立小孔内,保持小孔恰好处于充满状态。而后,将激光探头直接对准样品,启动激光器收集拉曼光谱信号,收集到的光谱信号记录为文本文档,使用软件绘制得到样品的拉曼光谱图。

4. 数据处理

拉曼光谱数据采集后使用美国恩威光电公司(Enwave Optronics)的 SLSR Reader V8.3.9 软件进行基线校正,校正后数据采用小波软阈值降噪法实施降噪,降噪后数据使用 mapminmax 函数进行归一化处理。主成分分析处理使用 princomp 函数,可有效降低数据维度,提高运算效率,选取累计贡献率 95% 以上

---

[1] 张正勇,岳彤彤,马杰,等.基于拉曼光谱与 k 最近邻算法的酸奶鉴别[J].分析试验室,2019,38(5):553-557.
[2] 王海燕,等.食药质量安全检测技术研究[M].北京:科学出版社,2023:1-215.

的主成分。分类识别算法使用 k 近邻算法,酸奶样品的 80% 样本用作训练集,余下的 20% 样本数用作测试集。数据运算分析及绘图平台:MATLAB R2016a。

6.1.2 结果与讨论

1. 酸奶拉曼光谱分析

三种品牌酸奶源自不同厂家,但口味较为相似,均为原味口味,外观均为白色黏稠状,人眼不能实现品牌鉴别。使用拉曼光谱仪采集酸奶的拉曼光谱数据,如图 6-1 所示。截至目前,关于酸奶的拉曼光谱分析鲜有报道,结合已有乳制品相关的拉曼光谱文献,对酸奶的主要拉曼光谱峰进行了归属分析,如表 6-1 所示[1][2]。进一步谱图分析发现,1 460 cm^{-1} 和 1 012 cm^{-1} 这两个是酸奶样品中强度排名前两位的拉曼光谱峰,分别对应于糖类和脂肪分子 CH$_2$ 变形振动和蛋白质(阿斯巴甜)苯丙氨酸的苯环振动(环内 C—C 对称伸缩),结合表 6-1 其他位置拉曼光谱出峰,可以明显看出本实验获得的酸奶拉曼光谱峰的贡献来源主要是蛋白质、糖类、脂肪和阿斯巴甜分子,这是酸奶的主要营养成分和甜味剂。同时,与本研究小组前期乳粉拉曼光谱分析有很大不同[3],在乳粉拉曼光谱表征中,最高峰只有 1 460 cm^{-1} 处 1 个峰,同时,蛋白质的拉曼光谱峰仅在 1 012 cm^{-1} 处有 1 个很小的峰,以及在 1 664 cm^{-1} 处有部分源自蛋白质的酰胺 I 键 C=O 伸缩振动贡献。在本实验中,发现蛋白质的拉曼光谱峰除了 1 012 cm^{-1} 和 1 664 cm^{-1} 处 2 个峰以外,还有 1 612 cm^{-1} 处蛋白质苯丙氨酸的苯环振动(环内 C—C 伸缩)以及 1 564 cm^{-1} 处的蛋白质酰胺 II 键的 C—N 伸缩振动、N—H 变形振动。造成这种现象的原因可能在于,酸奶在加工过程中添加了乳酸菌,经过乳酸菌发酵使得原料乳中的蛋白质等转变为氨基酸或多肽,以及酸奶中含有的阿斯巴甜分子,于是获得了图示的拉曼光谱信号。

---

[1] 刘文涵,杨未,张丹.苯丙氨酸银溶胶表面增强拉曼光谱的研究[J].光谱学与光谱分析,2008,28(2):343-346.

[2] Mazurek S, Szostak R, Czaja T, et al. Analysis of milk by FT-Raman spectroscopy[J]. Talanta, 2015, 138: 285-289.

[3] 张正勇,沙敏,刘军,等.基于高通量拉曼光谱的奶粉鉴别技术研究[J].中国乳品工业,2017,45(6):49-51.

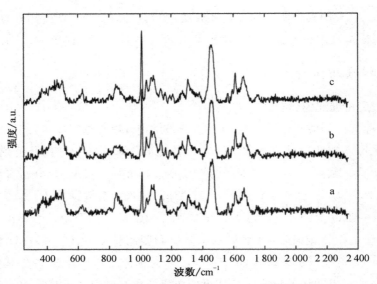

图 6-1　不同品牌 P1(a)、P2(b)和 P3(c)酸奶拉曼光谱图

表 6-1　酸奶拉曼光谱峰归属表

波数/cm^{-1}	归属
1 753	C=O 伸缩振动,主要可能来自脂肪有关的酯基
1 664	C=O 伸缩振动和 C=C 伸缩振动,其中 C=O 伸缩振动可能主要来自蛋白质的酰胺Ⅰ键,C=C 伸缩振动主要来自不饱和脂肪酸
1 612	苯丙氨酸的苯环振动,主要来自蛋白质、阿斯巴甜
1 564	N—H 变形振动,酰胺Ⅱ键的 C—N 伸缩振动,可能主要来自蛋白质
1 460	CH_2 变形振动,可能主要来自糖类和脂肪分子
1 312	CH_2 扭曲振动,可能主要来自脂肪酸
1 270	CH_2 扭曲振动,可能主要来自糖类
1 137	C—C 伸缩振动、C—O 伸缩振动以及 C—O—H 变形振动,主要可能来自糖类
1 081	C—C 伸缩振动、C—O 伸缩振动以及 C—O—H 变形振动,主要可能来自糖类
1 012	苯丙氨酸的苯环振动,主要来自蛋白质、阿斯巴甜
930	C—O—C 变形振动、C—O—H 变形振动和 C—O 伸缩振动,主要可能来自糖类
847	C—C—H 变形振动和 C—O—C 变形振动,主要可能来自糖类
627	C—C—O 变形振动
499	C—C—C 变形振动、C—O 扭曲振动

2. 酸奶拉曼光谱预处理分析

由图 6-1 可以看出,三种品牌酸奶的拉曼光谱图在出峰位置上具有较高的相似性,说明酸奶组分成分较为一致,不过在峰形状、峰间比值上有些许差异,因此可用模式识别算法予以测试分析,本节选用 k 近邻算法。在进行 k 近邻算法判别分析时,研究发现可对酸奶拉曼光谱进行预处理,以提高判别准确性和运算效率。首先,仔细观察酸奶拉曼光谱(图 6-1),可以发现原始光谱图中存在噪声,噪声是仪器设备采集信号时产生的随机信号,可能会对判别模型产生干扰。因此本节选用小波软阈值降噪方法进行谱图降噪预处理,基本步骤是针对谱图信号进行小波分解,选择 1 个小波并确定 1 个分解层次 N,对分解得到的小波系数进行阈值处理,最后根据小波分解的第 N 层的系数和高频系数进行信号的重构,得到平滑去噪后的谱图数据。如图 6-2 展示了基于 bior2.4 小波基的谱图降噪分析,图 6-2A 为 P1 品牌酸奶的拉曼光谱原始信号,图 6-2B 为 bior2.4 小波分解的第一层低频系数,图 6-2C 为 bior2.4 小波分解的第一层高频系数,可以看出光谱噪声主要集中在高频中,图 6-2D 为 bior2.4 小波降噪后重构的光谱信号,可以较明显地看出,光谱谱线变得更为平滑。目前常用的小波基有 bior、coif、db 和 sym 系列,为此,比较了在固定判别模型其他参数的条件下,其他参数是主成分选取前 40 个,k 取 1,马氏距离,分解层数 $N=3$,仅改变小波基,又由于测试集、训练集的选取使用的是随机选取法,因此各取 5 次试验结果,计算平均识别率的变化情况比较小波基降噪效果,如表 6-2 所示。在同样测试条件下,未做小波降噪的平均识别率是 93.18%,可以看出经过小波降噪,可以明显提高识别率,在 bior2.4 小波基条件下可以达到 99.70%。主成分分析是一种线性特征提取、降低数据维度、提高运算效率的数据预处理方法,如图 6-3、图 6-4 所示,拉曼光谱原有维度为 2 090 维,经过主成分提取,仅需使用 40 个提取的主成分,即可以达到原有数据信息的累计贡献率 95% 以上。在其他实验参数不变的条件下,其他参数是主成分选取前 40 个,k 取 1,马氏距离,分解层数 $N=3$,bior2.4 小波基,经过计算发现未做主成分降维的 5 次平均识别率为 99.38%,平均实验用时 0.887 9 s,使用主成分降维的 5 次平均识别率为 99.70%,平均实验用时 0.783 8 s,实验识别率均可达到 90% 以上,但实验用时减少了 0.104 1 s,运算效率提高 10% 以上。

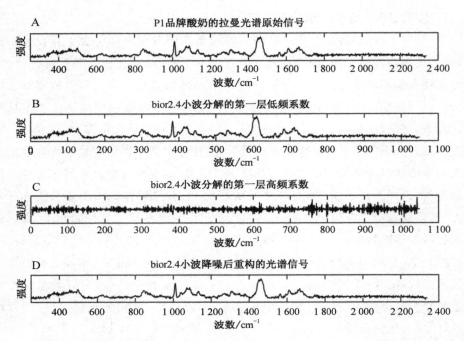

图 6-2 P1 品牌酸奶的拉曼光谱图(A),bior2.4 小波分解的第一层低频系数(B),bior 降噪后 2.4 小波分解的第一层高频系数(C),bior2.4 小波降噪后重构的光谱信号(D)

表 6-2 小波降噪方法比较结果

降噪方法	bior1.1	bior1.5	bior2.2	bior2.4	bior3.1	coif1	coif2	coif3	coif4	coif5
平均识别率	96.50%	98.82%	97.72%	99.70%	94.21%	99.13%	99.50%	98.95%	99.37%	99.25%
降噪方法	db1	db2	db3	db4	db5	sym1	sym2	sym3	sym4	sym5
平均识别率	96.83%	98.98%	99.17%	99.35%	98.02%	96.75%	99.15%	99.07%	99.07%	98.10%

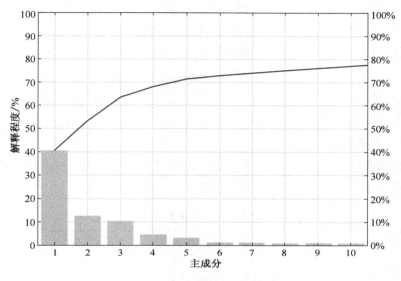

图 6-3 主成分分析帕累托图

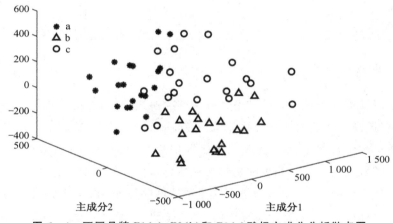

图 6-4 不同品牌 P1(a)、P2(b) 和 P3(c) 酸奶主成分分析散点图

3. 酸奶分类识别分析

  k 近邻算法是模式识别中经典的分类识别方法,其核心思想是根据样本在特征空间中的 k 个最近邻样本中大多数属于某个类别,即判断该样本也属于该类别,因此,k 是 k 近邻算法关键参数之一[①]。同时,计算类别时,常常使用马氏距离

---

  ① 王海燕,宋超,刘军,等.基于拉曼光谱—模式识别方法对奶粉进行真伪鉴别和掺伪分析[J].光谱学与光谱分析,2017,37(1):124-128.

或欧氏距离,为此,比较了在固定判别模型其他参数的条件下,其他参数是主成分选取前40个,分解层数 $N=3$,bior2.4小波基,不断改变 k 的取值,讨论了分别在马氏距离、欧氏距离条件下,判别模型的5次平均识别率,结果如表6-3所示。依据表6-3可以看出,$k=1$ 时,平均识别率值最大,接近100%,随着 k 值的增加,平均识别率均有下降,说明在本实验体系条件下,k 取值为1是最适宜的。比较马氏距离和欧氏距离结果,平均识别率差异不明显,进一步结合已有文献报道认为,欧氏距离是根据两点间对应坐标值之差的平方和再开方,而马氏距离是用协方差阵来把距离标准化后转化为无量纲的量来作为样本空间中两点的距离,判别结果更为合理,因此,本节最终优化选择使用马氏距离实现kNN判别[①]。

表6-3 k 取值优化表

k 取值	$k=1$	$k=3$	$k=5$	$k=7$	$k=9$	$k=11$	$k=13$	$k=15$	$k=17$	$k=19$
平均识别率 (马氏距离)	99.70%	98.85%	98.77%	96.31%	94.83%	92.60%	91.90%	90.38%	89.02%	88.18%
平均识别率 (欧氏距离)	99.65%	98.67%	98.65%	96.37%	94.67%	92.33%	91.50%	89.73%	89.18%	88.18%

6.1.3 小结

本节研究了基于拉曼光谱的市售品牌酸奶鉴别分析,通过采集与分析酸奶的拉曼光谱,展现了拉曼光谱快捷的采集优势,每个样品拉曼光谱采集仅需100 s,光谱解析揭示了酸奶丰富的营养成分表征信息,尤其是氨基酸、多肽含量丰富。尽管实验随机选择的三种品牌酸奶拉曼光谱具有极高的相似性,但经过谱图预处理和参数优化筛选后,得到适用于实验体系的最优条件:小波降噪(分解层数 $N=3$,bior2.4小波基),主成分分析选取前40个主成分(累计贡献率超过95%),k 近邻算法($k=1$,马氏距离),建立了以优化的 k 近邻算法为识别手段的快速识别方法。研究结果显示在最优参数条件下,平均识别率达到99.70%,智能判别时间仅需不足1 s。由此,建立了一种快速、准确判别酸奶品牌质量的鉴别方法,并可为其他食品品质控制智能技术研发提供借鉴。

---

① 梅江元.基于马氏距离的度量学习算法研究及应用[D].哈尔滨:哈尔滨工业大学,2016.

6.2 基于支持向量机算法的乳制品分类识别技术研究

乳制品的质量安全问题与每个人息息相关,其质量安全风险主要源于两个方面:一是有害物质或非法添加物,二是假冒伪劣、以次充好,比较典型的案例有2008年中国奶制品污染事件、2016年的假奶粉事件等[①]。为杜绝此类问题,监管部门制定并实施了多项产品标准、检测标准,如《食品安全国家标准 发酵乳》(GB 19302—2010),《食品安全国家标准 灭菌乳》(GB 25190—2010)、《食品安全国家标准 巴氏杀菌乳》(GB 19645—2010)等,规定了合格乳制品的原料要求、感官要求、理化指标、污染物限量、真菌毒素限量、微生物限量和其他要求,以及与各项指标要求对应的常规检测方法。不过,现有方法也存在着一定的挑战性,主要表现在感官检验与品评者的身体、技能、经验密切相关,有一定的主观性;常规的仪器成分检测法定性、定量分析准确,但一般需要前处理步骤和专业技术人员,较为耗时耗力;部分假冒产品实为低端产品冒充质优产品,其指标可能符合国家标准的基本要求,造成高效识别困难[②③]。此外,近年来,乳制品产量、社会需求量均逐年递增,迫切需要发展快速、高效的识别方法。

目前,现有的快速检测方法研究主要集中于比色法、胶体金试纸条法以及计算机辅助识别技术等[④⑤]。较之前两种方法,计算机辅助识别技术具备快速准确、客观、信息利用率高等多种优势,成为快速检测方法研发的热点。因此,本节首先采集了不同品牌的巴氏杀菌热处理风味发酵乳的质量特性数据拉曼光谱,随后运用支持向量机模式分类算法并对该方法进行参数优化,实现了乳制品快速分类识别。该方法具有多种优势,如拉曼光谱采集速度快、对操作人员要求低、无需样品前处理,支持向量机算法运算速度快、数据处理在 10 s 内即可完成等,为乳制品质量安全监管提供了技术参考,具备一定的潜在应用价值[⑥]。

---

① 剧柠,胡婕.光谱技术在乳及乳制品研究中的应用进展[J].食品与机械,2019,35(1):232-236.
② 郭文辉,袁彩霞,洪霞,等.乳制品中氰化物的快速检测[J].中国乳品工业,2019,47(2):61-64.
③ 张群.乳制品中抗生素的荧光快速检测技术研究及应用[J].食品与生物技术学报,2018,37(12):1336.
④ 石彬,李咏富,吴远根.氯化血红素比色法检测乳制品中土霉素[J].中国酿造,2018,37(7):168-172.
⑤ 赵小旭,柳家鹏,柴艳兵,等.胶体金免疫层析法快速检测乳制品中重金属离子铅[J].粮食科技与经济,2018,43(3):51-54.
⑥ 王梓笛,李双妹,尹延东,等.基于支持向量机算法的乳制品分类识别技术研究[J].粮食科技与经济,2020,45(3):104-107.

6.2.1 材料与方法

1. 材料

实验用巴氏杀菌热处理风味发酵乳购置于南京苏果超市,均为原味口味,选取三个品牌,莫斯利安巴氏热处理风味酸奶产品标记为品牌 P1,安慕希巴氏热加工风味酸奶产品标记为品牌 P2,纯甄巴氏杀菌热处理风味酸奶产品标记为品牌 P3。每种品牌随机采样 30 个样品,共计 90 个样品。

2. 仪器与设备

便携式激光拉曼光谱仪,光谱仪型号:Prott-ezRaman-D3,厂家:美国恩威光电公司(Enwave Optronics),激光波长为 785nm,激光最大功率约为 450 mW;电荷耦合器件检测器,温度控制在 -85 ℃ 左右,照射时间为 2.5 min,扫描次数为 1 次,光谱范围为 $250 \sim 2\,000\ cm^{-1}$,光谱分辨率为 $1\ cm^{-1}$;96 孔板:美国康宁公司(Corning Incorporated)。

3. 拉曼光谱图采集方法

取适量液态发酵乳直接上样,置于 96 孔板的独立小孔内,保持小孔恰好处于充满状态。之后,利用激光拉曼光谱仪探头直接照射样品,测试获取信号即为发酵乳的拉曼光谱数据。

4. 数据处理

采集样品的拉曼光谱数据后,使用 SLSR Reader V8.3.9 软件进行基线校正,校正后的光谱数据采用小波软阈值降噪法(wden 函数)实施噪音消除处理,然后使用 mapminmax 函数对光谱数据进行归一化处理,归一化至[-1,1]区间,使用 princomp 函数进行主成分分析。本次实验选取 80 个主成分,累计贡献率达到 99.2%。支持向量机分类识别算法使用 LIBSVM 工具箱[台湾大学林智仁(Lin Chih-Jen)教授等开发设计]实现算法的运算,使用 randperm 函数实现发酵乳样品随机抽样,以总样品数的 83%(每个品牌样品 25 个,共计 75 个样品)构建训练集,以剩下的 17%(每个品牌样品 5 个,共计 15 个样品)样品数据作为测试集。上述函数的运算、支持向量机算法及绘图使用 MATLAB 软件实现,版本为 R2016a。

6.2.2 结果与讨论

1. 发酵乳拉曼光谱分析

实验采集了三种不同品牌的巴氏杀菌热处理风味发酵乳,均呈液态黏稠白色试样,样品的拉曼光谱如图 6-5 所示,每个品牌随机选取了 10 个样品的拉曼光谱,其相互间保持了较高的一致性。此外,不同品牌拉曼光谱谱图之间,同样表现出较高的相似性,仅凭人眼难以实现有效辨别。拉曼光谱是一种基于样品分子与辐射光作用的散射光谱,适用于表征分子振动模式,对图6-5中出现的各主要拉曼光谱峰进行了信息归属,如表6-4所示,呈现出样品中的糖类、脂类、蛋白质等营养成分的拉曼光谱特征,如1 755 cm^{-1}主要源自于脂肪分子的C=O伸缩振动。各样品谱图间呈现出微小的峰形状、峰位置、峰高、峰比值差异[11]。据此可知,实验获得了发酵乳的拉曼光谱这一质量特性数据,为后续计算机识别模型的研究提供了必要的数据输入。

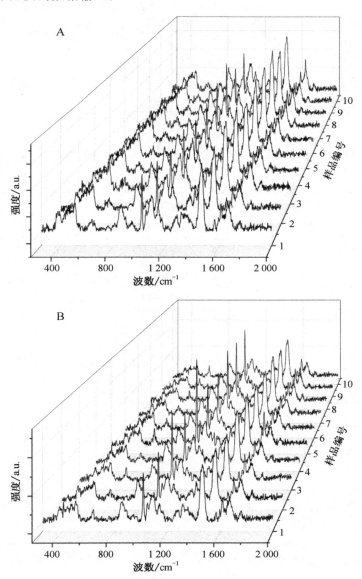

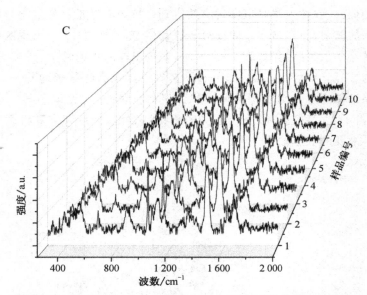

图 6-5 不同品牌 P1(A)、P2(B) 和 P3(C) 发酵乳的拉曼光谱图

表 6-4 发酵乳的拉曼光谱峰归属

波数/cm^{-1}	归属
1 755	$\nu(C=O)_{酯}$
1 666	$\nu(C=O)$酰胺Ⅰ,$\nu(C=C)$
1 615	$\nu(C-C)_{苯环振动(苯丙氨酸)}$
1 569	$\delta(N-H)$,$\nu(C-N)$酰胺Ⅱ
1 465	$\delta(CH_2)$
1 316	$\tau(CH_2)$
1 136	$\nu(C-O)+\nu(C-C)+\delta(C-O-H)$
1 090	$\nu(C-O)+\nu(C-C)+\delta(C-O-H)$
1 043	$\nu(C-O)+\nu(C-C)+\delta(C-O-H)$
1 014	$\nu(C-C)_{苯环振动(苯丙氨酸)}$
851	$\delta(C-C-H)+\delta(C-O-C)$
634	$\delta(C-C-O)$
503	葡萄糖
378	乳糖

注:ν 为伸缩振动,δ 为变形振动,τ 为扭曲振动,γ 为面外弯曲振动。

2. 发酵乳拉曼光谱的数据预处理

采集的样品拉曼光谱数据易出现噪声干扰,结合已有报道[1][2],对光谱数据进行了小波软阈值降噪,优化选用 db1 小波基,有效降低了光谱噪声的影响。为消除光谱数据量纲对分类模型的影响,对光谱数据进行了归一化处理,将数据强度值校正到 [-1,1] 区间。为提高光谱数据分类模型运算效率,对数据进行了主成分降维,每个样品原始拉曼光谱数据有 1751 个数据点,经主成分降维后,结果显示第 1 主成分可以达到原始数据的 43.4% 的解释程度,第 2 主成分可以达到原始数据的 9.9% 的解释程度,第 3 主成分可以达到原始数据的 5.0% 的解释程度。实验选用 80 个主成分,可代表原有信息的累计贡献率达到 99.2%,既保留了原始光谱数据的主要信息又提高了运算效率。主成分分析结果如图 6-6 所示,选取了拉曼光谱原始数据提取后的第 1 和第 2 主成分,可以看出,同品牌样品间倾向于聚集,不同品牌样品间倾向于分离,揭示出尽管发酵乳制品的拉曼光谱原始数据具有较高的相似性,但是同品牌样品间数据统计的相似性要高于不同品牌样品间的相似性。

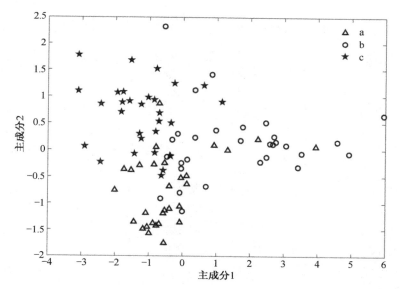

图 6-6　不同品牌 P1(a)、P2(b) 和 P3(c) 发酵乳主成分分析图

---

① Fang S Y, Wu S Y, Chen Z, et al. Recent progress and applications of Raman spectrum denoising algorithms in chemical and biological analyses: A review[J]. TrAC Trends in Analytical Chemistry, 2024, 172: 117578.

② Ehrentreich F, Sümmchen L. Spike removal and denoising of Raman spectra by wavelet transform methods[J]. Analytical Chemistry, 2001, 73(17): 4364-4373.

3. 基于支持向量机算法的发酵乳判别分析

支持向量机算法是基于统计学习理论建立起来的一种模式分类识别方法,其核心是建立一个分类超平面作为决策曲面,该曲面对待分类的不同种类样本进行正确分类,并使分类后的样本点距离该分类超平面最远,支持向量决定了这一最优分类界面[1][2][3]。本节使用 LIBSVM 工具箱,通过一对一法来解决多分类问题,基本思路是在任意两类样品之间设计一个两分类支持向量机,当对一个未知样本进行分类时,得票最多的类别即为该未知样本的类别。如图 6-7 所示,图中展示了以两个不同品牌的发酵乳制品光谱数据为输入,运用 svmplot 函数绘制出通过计算获得的样品间支持向量分布情况,A 图显示了训练集分类情况,B 图显示了测试集分类情况,由图显示出通过训练集获得支持向量构建的决策曲面为非线性曲面,能够实现样品分类识别。

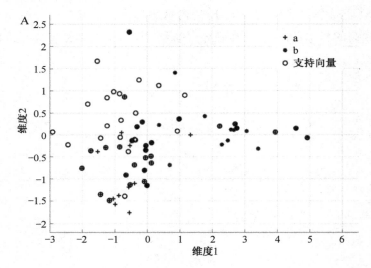

---

① 李志豪,沈俊,边瑞华,等. 机器学习算法用于公安一线拉曼实际样本采样学习及其准确度比较[J]. 光谱学与光谱分析,2019,39(7):2171-2175.

② 陈思雨,张舒慧,张纡,等. 基于共聚焦拉曼光谱技术的苹果轻微损伤早期判别分析[J]. 光谱学与光谱分析,2018,38(2):430-435.

③ Du L J, Lu W Y, Cai Z J, et al. Rapid detection of milk adulteration using intact protein flow injection mass spectrometric fingerprints combined with chemometrics[J]. Food Chemistry, 2018, 240: 573-578.

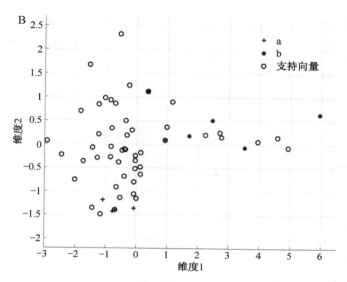

图 6-7　基于支持向量机的训练集(A),测试集(B)发酵乳分类可视化图

注:a 代表品牌 P1,b 代表品牌 P2。

支持向量机模式分类识别算法的实施步骤如下:首先,将经过降噪、归一化、主成分提取后的训练集样品拉曼光谱数据导入支持向量机算法;而后,选择一个恰当的核函数,常用的核函数有线性核函数、多项式核函数、径向基核函数等,本节选用应用最为广泛的径向基核函数,将样本特征从较低维输入空间映射到高维特征空间;最后,根据由优化问题求解而来的支持向量得到相应的决策函数,本节使用 svmtrain 函数及训练集进行模型构建。核函数参数 g 及惩罚系数 c 是影响模型识别效果的关键参数,两者的选择决定了分类模型的识别精度,核函数参数 g 决定了输入空间映射到高维特征空间的方式,惩罚系数 c 决定了平衡训练误差和模型复杂度[1][2]。本节采用 SVMcgForClass 函数进行网格参数寻优,设置 5 折交叉验证法($K=5$),核函数参数 g 和惩罚系数 c 的参数寻优条件是 $g_{min}=-10$,$g_{max}=10$,$c_{min}=-10$,$c_{max}=10$,搜索范围是 $[2^{-10},2^{10}]$,步进值均为 0.5,最终获得了最优核函数 g 为 0.022 097,惩罚系数 c 为 32,分类模型识别率达到最大。分类识别结果如图 6-8 所示,在优化条件下,测试集的分类模型最高识别结果可达到

---

[1] 张文雅,范雨强,韩华,等.基于交叉验证网格寻优支持向量机的产品销售预测[J].计算机系统应用,2019,28(5):1-9.

[2] Zhang L G, Zhang X, Ni L J, et al. Rapid identification of adulterated cow milk by non-linear pattern recognition methods based on near infrared spectroscopy[J]. Food Chemistry, 2014, 145:342-348.

100%的准确率。

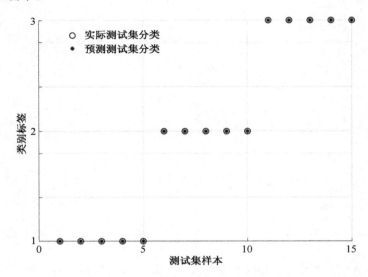

图6-8 基于支持向量机的发酵乳测试集分类结果图

注：类别1代表品牌P1、类别2代表品牌P2、类别3代表品牌P3发酵乳制品。

6.2.3 小结

实验以巴氏杀菌热处理风味发酵乳的品牌分类识别为例，研究探讨了以拉曼光谱数据为输入，以支持向量机算法为判别手段的乳制品计算机识别技术。该技术展现出拉曼光谱快捷方便的采集优势，每个样品的拉曼光谱数据采集仅需2.5分钟，操作简单，可直接上样测试。针对实验样品及拉曼光谱图表现出较高的相似性、人眼难以判别的情况，发展了面向对象的支持向量机判别方法，经过谱图预处理和参数优化筛选后，得到适用于分类体系的优化条件：小波软阈值降噪（db1小波基，分解层数$N=3$），主成分分析选取前80个主成分（累计贡献率达99%以上），支持向量机（径向基核函数，核函数参数g为0.022 097，惩罚系数c为32），据此，实现了乳制品快速模式分类，识别所需时间不足10 s。

6.3 基于拉曼光谱与随机森林算法相结合的豆乳粉品牌识别

食品质量控制主要集中在两个方面：一是特定成分的识别，包括营养成分和有害成分,二是类别判别①。拉曼光谱作为一种高效的分子振动光谱表征技术,引起了研究者的广泛兴趣。例如 Schulmerich 等报道,使用透射拉曼光谱结合偏最小二乘法可以预测单个大豆的油和蛋白质含量②。Yin 等研究了大豆蛋白中甘氨酸的分离和拉曼光谱分析③。拉曼光谱可以实现样品的无损检测,无需样品预处理,而且拉曼光谱仪方便携带,有利于现场检测④⑤。目前,基于拉曼光谱的食品成分检测方法的报道越来越多,特别是表面增强拉曼光谱,它通过特定的金和银纳米材料以及分子相互作用或分子修饰的识别,实现对某些特定分子的定性和定量分析⑥。Xi 等最近报道了一种基于表面增强拉曼光谱与金纳米粒子偶联的横向流免疫条带测试,该测试可以实现大豆过敏原 β-伴球蛋白的快速检测⑦,利用上述方法实现了大豆中某些特征分子的拉曼光谱分析。

近年来,食品类别判别也引起了研究者的广泛关注。例如,时常有媒体报道市场上出现了假冒产品,这些产品对消费者的健康存在风险。然而,对于分类判别问题来说,仅通过目前的特定成分方法通常很难获得准确的判别结果。近年来,将机器学习算法与光谱表征技术相结合用于食品判别和分析的报道迅速增加⑧。例如,Barros 等利用特征拉曼光谱带和普通最小二乘法实现了对特级初榨橄榄油中大豆油的定量检测⑨。随机森林算法是机器学习中的一种集成学习方

---

① Battisti I, Ebinezer L B, Lomolino G, et al. Protein profile of commercial soybean milks analyzed by label-free quantitative proteomics[J]. Food Chemistry, 2021, 352: 129299.

② Schulmerich M V, Walsh M J, Gelber M K, et al. Protein and oil composition predictions of single soybeans by transmission Raman spectroscopy[J]. Journal of Agricultural and Food Chemistry, 2012, 60(33): 8097-8102.

③ Yin H C, Huang J, Zhang H R. Study on isolation and Raman spectroscopy of glycinin in soybean protein[J]. Grain & Oil Science and Technology, 2018, 1(2): 72-76.

④ de Oliveira Mendes T, Manzolli Rodrigues B V, Simas Porto B L, et al. Raman Spectroscopy as a fast tool for whey quantification in raw milk[J]. Vibrational Spectroscopy, 2020, 111: 103150.

⑤ Vasafi P S, Hinrichs J, Hitzmann B. Establishing a novel procedure to detect deviations from standard milk processing by using online Raman spectroscopy[J]. Food Control, 2022, 131: 108442.

⑥ Li L M, Chin W S. Rapid and sensitive SERS detection of melamine in milk using Ag nanocube array substrate coupled with multivariate analysis[J]. Food Chemistry, 2021, 357: 129717.

⑦ Xi J, Yu Q R. The development of lateral flow immunoassay strip tests based on surface enhanced Raman spectroscopy coupled with gold nanoparticles for the rapid detection of soybean allergen β-conglycinin[J]. Spectrochimica Acta Part A, Molecular and Biomolecular Spectroscopy, 2020, 241: 118640.

⑧ Zhao H F, Zhan Y L, Xu Z, et al. The application of machine-learning and Raman spectroscopy for the rapid detection of edible oils type and adulteration[J]. Food Chemistry, 2022, 373(Pt B): 131471.

⑨ Barros I H A S, Santos L P, Filgueiras P R, et al. Design experiments to detect and quantify soybean oil in extra virgin olive oil using portable Raman spectroscopy[J]. Vibrational Spectroscopy, 2021, 116: 103294.

法,它是一个包含多个决策树的分类器。根据每个决策树的输出类别的模式来确定最终的分类结果,具有操作精度高、速度快的优点。例如,采用随机森林算法和红外光谱法来区分不同地理来源的三七[①]。然而,在大豆乳制品中仍然缺乏相关的应用研究案例。因此,针对豆乳粉的品牌判别需求,本节收集了不同品牌豆乳粉的拉曼光谱数据,研究了基于随机森林算法和拉曼光谱处理相结合的判别技术[②]。

6.3.1 实验部分

1. 材料

实验用豆乳粉均购置于南京苏果超市,其中,维维豆乳粉标记为品牌 P1,永和豆乳粉标记为品牌 P2,润之家豆乳粉标记为品牌 P3。每个品牌有 60 个样品,总共 180 个样品。

2. 仪器与设备

将适当量的样品填充到 96 孔板的每个小孔中,然后使用便携式激光拉曼光谱仪获得每个豆乳粉样品的拉曼光谱。光谱仪型号:Prott-ezRaman-D3,厂家:美国恩威光电公司(Enwave Optronics),采集参数包括 785 nm 的激光波长、450 mW 的激光功率、40 s 的照射时间、-85 ℃ 的电荷耦合器件温度和 $250\sim 2\,000\ cm^{-1}$ 的光谱范围,光谱分辨率为 $1\ cm^{-1}$。拉曼光谱采集过程中,样品无需进行任何额外的物理或化学预处理。96 孔板:美国康宁公司(Corning Incorporated)。

3. 数据采集与处理

通过光谱仪自带软件 SLSR Reader V8.3.9 对获得的各样品拉曼光谱信号进行基线校正。拉曼光谱数据的处理和分析包括小波去噪、归一化、主成分分析、求导和随机森林算法等,并使用 MATLAB 平台(美国 MathWorks 公司)进行计算。

6.3.2 结果和讨论

1. 豆乳粉的拉曼光谱分析

不同品牌豆乳粉的拉曼光谱如图 6-9 所示。参考已有的相关文献,可以尝

---

① Li Y, Zhang J Y, Wang Y Z. FT-MIR and NIR spectral data fusion: A synergetic strategy for the geographical traceability of Panax notoginseng[J]. Analytical and Bioanalytical Chemistry, 2018, 410(1): 91-103.

② Zhang Z Y, Shi X J, Zhao Y J, et al. Brand Identification of Soybean Milk Powder based on Raman Spectroscopy Combined with Random Forest Algorithm[J]. Journal of Analytical Chemistry, 2022, 77(10): 1282-1286.

试对拉曼光谱主要谱峰进行归属分析[1][2][3][4]。1 674 cm^{-1}的波数主要归因于蛋白质酰胺Ⅰ的C═O伸缩振动和不饱和脂肪酸的C═C伸缩振动,1 470 cm^{-1}处的光谱源自脂肪和碳水化合物的CH$_2$变形振动。在1 355 cm^{-1}处拉曼峰主要是源自碳水化合物的C—O伸缩振动和C—O—H变形振动,在1 090 cm^{-1}和1 140 cm^{-1}处的拉曼峰可归属于碳水化合物的C—O伸缩振动、C—C伸缩振动和C—O—H变形振动。在1 019 cm^{-1}处出现的是一个较为特殊的拉曼峰,这是蛋白质苯丙氨酸的苯环呼吸伸缩振动。在250~1 000 cm^{-1}的范围内有一系列光谱峰,主要与碳水化合物分子有关,如945 cm^{-1}(C—O—C、C—O—H变形振动和C—O伸缩振动)、850 cm^{-1}(C—C—H和C—O—C变形振动)、768 cm^{-1}(C—C—O变形振动)、712 cm^{-1}(C—S伸缩振动)、640 cm^{-1}(C—C—O变形振动)、575 cm^{-1}(C—C—C变形振动和C—O扭曲振动)、521 cm^{-1}(葡萄糖)、433 cm^{-1}(C—C—C变形振动和C—O扭曲振动)以及345 cm^{-1}(乳糖),从而可以获得实验样品丰富的分子振动信息。

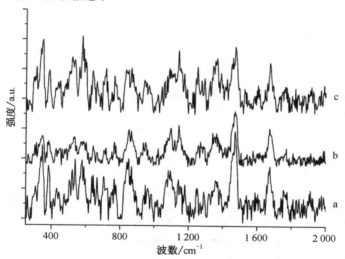

图 6-9　不同品牌 P1(a)、P2(b)和 P3(c)豆乳粉的拉曼光谱图

---

[1]　Wu L K, Wang L M, Qi B K, et al. 3D confocal Raman imaging of oil-rich emulsion from enzyme-assisted aqueous extraction of extruded soybean powder[J]. Food Chemistry, 2018, 249: 16-21.

[2]　El-Abassy R M, Eravuchira P J, Donfack P, et al. Fast determination of milk fat content using Raman spectroscopy[J]. Vibrational Spectroscopy, 2011, 56(1): 3-8.

[3]　Lee H, Cho B K, Kim M S, et al. Prediction of crude protein and oil content of soybeans using Raman spectroscopy[J]. Sensors and Actuators B: Chemical, 2013, 185: 694-700.

[4]　Jin H Q, Li H, Yin Z K, et al. Application of Raman spectroscopy in the rapid detection of waste cooking oil[J]. Food Chemistry, 2021, 362: 130191.

2. 豆乳粉拉曼光谱的数据处理分析

不同品牌豆乳粉的主要成分是蛋白质、碳水化合物和脂肪分子。根据 1 470 cm^{-1} 处各样品的强度值，计算了这些样品的拉曼光谱的相对标准偏差，结果表明，品牌 P1、P2 和 P3 的相对标准偏差分别为 14.6%、11.8% 和 9.0%，仅通过拉曼光谱的视觉比较很难实现各品牌的有效判别。因此，本节工作以拉曼光谱数据为输入，开展了基于随机森林算法的计算机辅助判别研究。首先，在没有任何光谱处理的情况下，150 个样本（每个品牌 50 个样本）用作训练集，其余 30 个样本（每个品牌 10 个样本）作为测试集，用于直接模型操作。识别准确率最高可达 96.7%（图 6-10）。这一结果初步表明，随机森林算法与拉曼光谱相结合可以实现对豆乳粉的有效判别。光谱处理分析有望进一步提取有效的光谱特征，减少冗余光谱信息的干扰，进一步提高模型的判别精度和运算效率[①]。

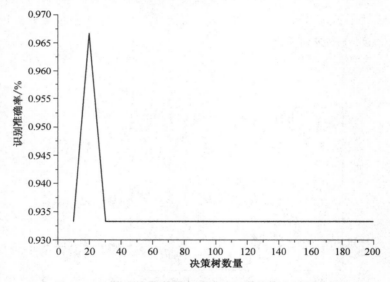

图 6-10 基于拉曼光谱全谱的随机森林算法判别结果

从豆乳粉样品的原始拉曼光谱可以看出，可能存在一定的信号噪声，如图 6-9 所示。小波函数可以用于降噪，其基本原理是对拉曼光谱信号进行小波分解，滤除高频噪声信号，得到重构后的去噪光谱信号，实验研究了"sym""db"和"coif"小

---

① Sha M，Gui D D，Zhang Z Y，et al. Evaluation of sample pretreatment method for geographic authentication of rice using Raman spectroscopy[J]. Journal of Food Measurement and Characterization，2019，13(3)：1705-1712.

波基[①]。识别结果如表6-5所示。在适当的小波去噪条件下,识别准确率有较大提高,表明滤波频谱噪声有利于判别模型,适合该实验系统的最佳小波条件为db2等。db2小波去噪后的拉曼光谱如图6-11所示,豆乳粉的拉曼光谱信号已被显著平滑。

表6-5 不同小波去噪后的豆乳粉判别结果

小波基	识别率/%	小波基	识别率/%	小波基	识别率/%
sym1	96.7	db1	93.3	coif1	100
sym2	96.7	db2	100	coif2	96.7
sym3	96.7	db3	100	coif3	96.7
sym4	96.7	db4	100	coif4	100
sym5	93.3	db5	100	coif5	96.7

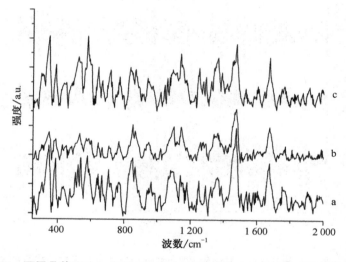

图6-11 不同品牌P1(a)、P2(b)和P3(c)豆乳粉在db2小波去噪后的拉曼光谱

---

① Zhao X Y, Liu G Y, Sui Y T, et al. Denoising method for Raman spectra with low signal-to-noise ratio based on feature extraction [J]. Spectrochimica Acta Part A, Molecular and Biomolecular Spectroscopy,2021,250: 119374.

在小波去噪之后，拉曼光谱的处理技术还可以包括导数、归一化和特征提取[1]。光谱求导被认为可进一步突出一些光谱细节。因此，对豆乳粉的拉曼光谱数据分别进行了一阶导数、二阶导数、三阶导数和四阶导数处理，如图6-12至图6-15所示。从这些图中可以看出，谱图信号变得复杂，直观上难以有效判别，将数据导入随机森林模型后，识别准确率没有显著提高，反而有所下降（表6-6），这进一步表明求导处理不适用于该实验系统。识别准确率下降的原因可能是求导运算也大大放大了噪声的干扰，反而影响了识别准确率[2]。归一化有助于消除不同样本的不同光谱信号量纲对判别模型的影响。归一化后，将所有豆乳粉样品的拉曼光谱强度设置到[-1,1]的区间，如图6-16所示。将数据导入随机森林模型后，识别准确率提高到100%。结果表明，归一化处理对于实验系统是有效的。

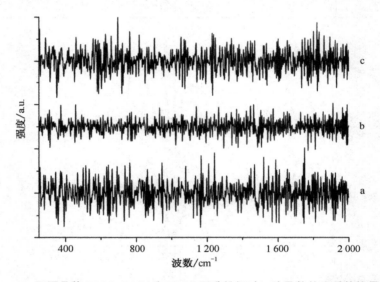

图6-12 不同品牌P1(a)、P2(b)和P3(c)豆乳粉经过一阶导数处理后的拉曼光谱

---

[1] Hu J Q, Zhang D, Zhao H T, et al. Intelligent spectral algorithm for pigments visualization, classification and identification based on Raman spectra[J]. Spectrochimica Acta Part A, Molecular and Biomolecular Spectroscopy, 2021, 250: 119390.

[2] 陈凤霞, 杨天伟, 李杰庆, 等. 基于偏最小二乘法判别分析与随机森林算法的牛肝菌种类鉴别[J]. 光谱学与光谱分析, 2022, 42(2): 549-554.

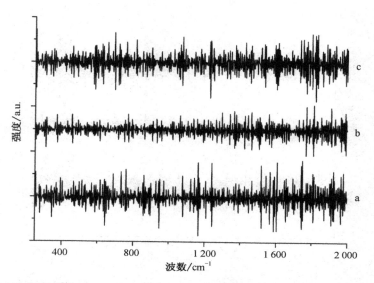

图 6-13　不同品牌 P1(a)、P2(b)和 P3(c)豆乳粉经过二阶导数处理后的拉曼光谱

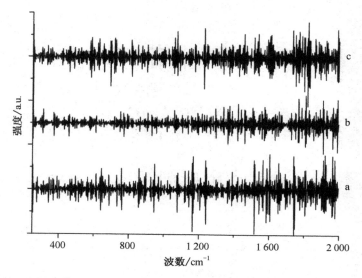

图 6-14　不同品牌 P1(a)、P2(b)和 P3(c)豆乳粉经过三阶导数处理后的拉曼光谱

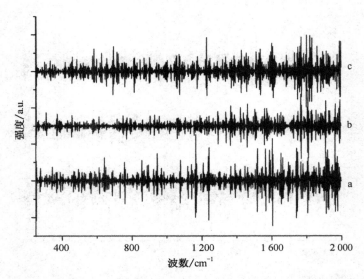

图 6-15 不同品牌 P1(a)、P2(b)和 P3(c)豆乳粉经过四阶导数处理后的拉曼光谱

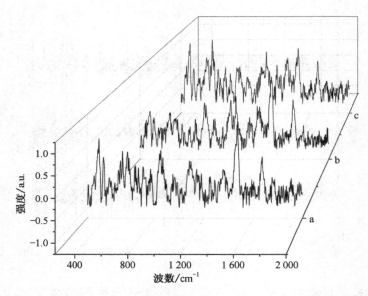

图 6-16 不同品牌 P1(a)、P2(b)和 P3(c)豆乳粉经过归一化处理后的拉曼光谱

表 6-6　不同光谱处理后的豆奶粉识别结果

光谱预处理	识别率/%
一阶导数	93.3
二阶导数	56.7
三阶导数	63.3
四阶导数	76.7
归一化	100

此外,主成分分析方法可以用于提取豆乳粉的拉曼光谱特征。作为一种常用的数据转换方法,它也可以有效地降低拉曼光谱数据的维数。经过主成分分析处理后,第一主成分可以代表原始光谱数据 34.9% 的信息,第二主成分可以代表原始光谱数据 13.1% 的信息,而第三主成分可以代表原始光谱数据 4.0% 的信息。如图 6-17 所示,仅 100 个主成分就可以代表原始频谱 1 751 个数据点 95% 以上的信息,可以有效提高模型的运行效率。主成分分析后的散点图如图 6-18 所示,可以直观地看出,不同品牌的豆乳粉样品趋向于分离,同一品牌的豆乳粉样品趋向于聚集。

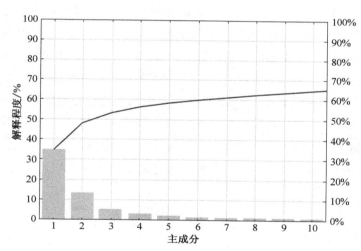

图 6-17　主成分分析结果的帕累托图

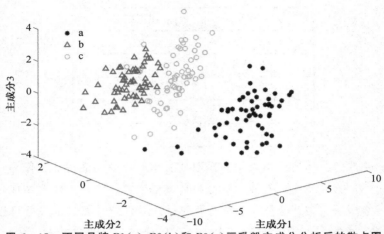

图 6-18 不同品牌 P1(a)、P2(b)和 P3(c)豆乳粉主成分分析后的散点图

3. 基于随机森林算法的豆乳粉快速判别

随机森林算法是一种包含多个决策树的分类器,然后根据大多数树的判别类别输出识别结果[1][2]。本节选取 150 个经过小波去噪、归一化和主成分处理的豆乳粉拉曼光谱数据(每个品牌 50 个)作为训练集,其余 30 个拉曼光谱数据作为测试集(每个品牌 10 个),随机循环操作 100 次,研究了不同决策树的数量对判别模型运算结果的影响,结果如图 6-19 所示。可以看出,当有 10 棵树和 20 棵树时,识别准确率分别为 86.7% 和 90%,当有 30 棵或更多棵树时,识别准确率达到并保持 100%。研究结果表明,决策树的数量对识别精度有一定的影响。对于这个实验系统,当有 30 个决策树时,可以获得更好的识别结果。此外,本节的实验研究还表明,随机森林算法在面临特定的分类和判别问题时,需要进行必要的光谱预处理和决策树优化,这可能也是该算法目前面临的挑战之一。

---

[1] Li M G, Xu Y Y, Men J, et al. Hybrid variable selection strategy coupled with random forest (RF) for quantitative analysis of methanol in methanol-gasoline via Raman spectroscopy[J]. Spectrochimica Acta Part A, Molecular and Biomolecular Spectroscopy, 2021, 251: 119430.

[2] de Santana F B, Mazivila S J, Gontijo L C, et al. Rapid discrimination between authentic and adulterated andiroba oil using FTIR-HATR spectroscopy and random forest[J]. Food Analytical Methods, 2018, 11(7): 1927-1935.

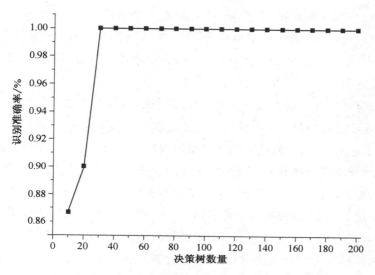

图 6-19 随机森林中决策树数量对拉曼光谱数据预处理后识别准确率的影响

6.3.3 小结

本节工作对豆乳粉的拉曼光谱进行了收集和分析,结果表明,样品中含有丰富的与碳水化合物、脂肪和蛋白质有关的分子振动信息。以拉曼光谱为数据输入,研究了小波去噪(db2 小波)、归一化([−1,1])和主成分分析(提取 100 个主成分)相结合的光谱处理方法,以及随机森林算法(30 棵树)。在最佳实验条件下,可以有效判别出不同品牌的豆乳粉样品。

6.4 基于拉曼光谱和概率神经网络的乳酪制品快速鉴别

近年来,乳制品的质量安全保障已成为消费者、企业和政府监管机构高度关注的热点问题,其风险主要来自非法添加、有毒有害物质和劣质产品[1]。现有被广泛研究和应用的检测方法包括成分分析,主要以色谱法、色谱—质谱法为代

---

[1] Wang J J, Wu Y, Wu Q H, et al. Highly sensitive detection of melamine in milk samples based on N-methylmesoporphyrin IX/G-quadruplex structure[J]. Microchemical Journal, 2020, 155: 104751.

表①。例如,检测三聚氰胺、双氰胺、黄曲霉毒素等物质,可以通过成分分析获得这些目标分子准确的定性和定量信息②③。然而,此类方法必须面对两个挑战。第一,它是一种常规的成分分离和分析检测,通常需要样品预处理,较为耗时费力。第二,近期时常有媒体报道,一些假冒产品其本质上是合格产品,并不会对人体造成直接伤害,只是被犯罪分子非法用来冒充优质产品,以差价的方式谋取非法利润。传统的成分分析法可能无法有效判别这类风险样品。

鉴于上述问题,快速检测方法在质量控制领域引起了广泛的关注,如比色法、试纸条法和计算机辅助判别法等④⑤。比色法主要是利用标记物和反应物之间的特定反应,或者标记物引起纳米材料(如胶体金纳米颗粒)的分散和聚集状态变化,然后通过颜色变化,进行目标分子判别⑥⑦。试纸条法主要利用免疫或竞争反应来引发或阻止胶体金纳米粒子在试纸条检测线上的聚集,检测线显示红色或无色,以判断目标分子的存在或不存在⑧⑨。计算机辅助判别法主要基于样品的光

---

① Zhou W E, Wu H Q, Wang Q, et al. Simultaneous determination of formononetin, biochanin A and their active metabolites in human breast milk, saliva and urine using salting-out assisted liquid-liquid extraction and ultra high performance liquid chromatography-electrospray ionization tandem mass spectrum[J]. Journal of Chromatography B, Analytical Technologies in the Biomedical and Life Sciences, 2020, 1145: 122108.

② Ramezani A M, Ahmadi R, Absalan G. Designing a sustainable mobile phase composition for melamine monitoring in milk samples based on micellar liquid chromatography and natural deep eutectic solvent[J]. Journal of Chromatography A, 2020, 1610: 460563.

③ Shuib N S, Makahleh A, Salhimi S M, et al. Determination of aflatoxin M_1 in milk and dairy products using high performance liquid chromatography-fluorescence with post column photochemical derivatization[J]. Journal of Chromatography A, 2017, 1510: 51 – 56.

④ Yan S, Lai X X, Du G R, et al. Identification of aminoglycoside antibiotics in milk matrix with a colorimetric sensor array and pattern recognition methods[J]. Analytica Chimica Acta, 2018, 1034: 153 – 160.

⑤ Na G Q, Hu X F, Yang J F, et al. Colloidal gold-based immunochromatographic strip assay for the rapid detection of bacitracin zinc in milk[J]. Food Chemistry, 2020, 327: 126879.

⑥ Luan Q, Gan N, Cao Y T, et al. Mimicking an enzyme-based colorimetric aptasensor for antibiotic residue detection in milk combining magnetic loop-DNA probes and CHA-assisted target recycling amplification[J]. Journal of Agricultural and Food Chemistry, 2017, 65(28): 5731 – 5740.

⑦ Ai K L, Liu Y L, Lu L H. Hydrogen-bonding recognition-induced color change of gold nanoparticles for visual detection of melamine in raw milk and infant formula[J]. Journal of the American Chemical Society, 2009, 131(27): 9496 – 9497.

⑧ Shi Q Q, Huang J, Sun Y N, et al. Utilization of a lateral flow colloidal gold immunoassay strip based on surface-enhanced Raman spectroscopy for ultrasensitive detection of antibiotics in milk[J]. Spectrochimica Acta Part A, Molecular and Biomolecular Spectroscopy, 2018, 197: 107 – 113.

⑨ Shin W R, Sekhon S S, Rhee S K, et al. Aptamer-based paper strip sensor for detecting Vibrio fischeri[J]. ACS Combinatorial Science, 2018, 20(5): 261 – 268.

谱、色谱、质谱等数据信息,并结合机器学习算法来进行运算判别[1][2]。由于光谱数据可以快速获得,并且包含丰富的样品分子信息,因此光谱法已成为快速检测技术研发领域的热点。光谱主要包括红外光谱、紫外光谱、荧光光谱和拉曼光谱等。红外光谱易受到水分子的干扰,紫外光谱法要求被测物体含有不饱和化合物,样品主要为液体,荧光光谱法要求被测物体含有发光结构。与上述光谱法相比,拉曼光谱法具有许多优点,如不受水分子的明显干扰,可以直接测试物体,可表征样品丰富的振动信息,且其设备具有便携性等[3][4],使之成为快速检测领域的研发关注点。例如,Mendes 等建立了一种基于振动光谱的乳脂定量方法[5],Teixeira 等从实验和理论水平评估了牛奶中 β-内酰胺类抗生素的检测[6],Nieuwoudt 等报道了一种利用偏最小二乘法和拉曼光谱法快速定量测定牛奶中三聚氰胺、尿素、硫酸铵、双氰胺和蔗糖的方法[7],冯彦婷等研究建立了一种基于纳米银颗粒团聚反应的表面增强拉曼光谱法,可用以牛乳中三聚氰胺的测定,检测限达到 0.08 mg/L[8]。不过,在乳制品领域,拉曼光谱与机器学习算法相结合的研究案例仍然相对较少,目前的报道主要集中在成分检测回归分析方面[9]。

  本节建立了一种基于拉曼光谱和概率神经网络的乳酪制品分类方法。乳酪

---

[1] Balan B J, Dhaulaniya A S, Jamwal R, et al. Application of Attenuated Total Reflectance-Fourier Transform Infrared (ATR-FTIR) spectroscopy coupled with chemometrics for detection and quantification of formalin in cow milk[J]. Vibrational Spectroscopy, 2020, 107: 103033.

[2] Lu W Y, Liu J, Gao B Y, et al. Technical note: Nontargeted detection of adulterated plant proteins in raw milk by UPLC-quadrupole time-of-flight mass spectrometric proteomics combined with chemometrics[J]. Journal of Dairy Science, 2017, 100(9): 6980-6986.

[3] Mazurek S, Szostak R, Czaja T, et al. Analysis of milk by FT-Raman spectroscopy[J]. Talanta, 2015, 138: 285-289.

[4] Karacaglar N N Y, Bulat T, Boyaci I H, et al. Raman spectroscopy coupled with chemometric methods for the discrimination of foreign fats and oils in cream and yogurt[J]. Journal of Food and Drug Analysis, 2019, 27(1): 101-110.

[5] Mendes T O, Junqueira G M A, Porto B L S, et al. Vibrational spectroscopy for milk fat quantification: Line shape analysis of the Raman and infrared spectra[J]. Journal of Raman Spectroscopy, 2016, 47(6): 692-698.

[6] Teixeira R C, Luiz L C, Junqueira G M A, et al. Detection of antibiotic residues in Cow's milk: A theoretical and experimental vibrational study[J]. Journal of Molecular Structure, 2020, 1215: 128221.

[7] Nieuwoudt M K, Holroyd S E, McGoverin C M, et al. Raman spectroscopy as an effective screening method for detecting adulteration of milk with small nitrogen-rich molecules and sucrose[J]. Journal of Dairy Science, 2016, 99(4): 2520-2536.

[8] 冯彦婷,林沛纯,谢慧凤,等.基于纳米银颗粒团聚反应的表面增强拉曼光谱法测定牛奶中三聚氰胺的含量[J].食品与发酵工业,2019,45(15):256-261.

[9] Alves da Rocha R, Paiva I M, Anjos V, et al. Quantification of whey in fluid milk using confocal Raman microscopy and artificial neural network[J]. Journal of Dairy Science, 2015, 98(6): 3559-3567.

制品含有丰富的营养成分,产品数量相对较少,价格较高,因此迫切需要开发一种快速、智能的鉴别方法[1]。新建立的方法主要包括以下优点。首先,传统的成分分析方法难以有效识别相似度高的不同品牌乳酪产品,样品分子信息可以通过拉曼光谱进行表征,结合概率神经网络分类算法有望实现样品鉴别。其次,实验样品的拉曼光谱信号采集简单、快速,不需要样品预处理,乳酪样品中的水分子不干扰测试,每个样品的数据采集时间仅为 100 s,概率神经网络算法运算时间小于 1 s。最后,拉曼光谱仪是便携式的,有利于现场检测。结合拉曼光谱的分子表征特性,该方法可以有效地实现实验样品的指纹图谱表征与分析,为乳制品鉴别提供了一个智能的客观评价技术支撑[2]。

6.4.1 实验部分

1. 样品和仪器

实验用乳酪制品均购自南京苏果超市,其中,妙可蓝多乳酪标记为品牌 P1,夏洛克乳酪标记为品牌 P2,伊利乳酪标记为品牌 P3。每个品牌随机抽取 25 个样品。

在 96 孔板中装入适量的乳酪制品。使用便携式激光拉曼光谱仪记录拉曼光谱,光谱仪型号:Prott-ezRaman-D3,厂家:美国恩威光电公司(Enwave Optronics)。激光的激发波长为 785 nm,激光功率约为 450 mW,照射时间为 100 s。光谱仪在 $250 \sim 2\,000\ cm^{-1}$ 的光谱范围内工作,光谱分辨率为 $1\ cm^{-1}$。

2. 数据处理

收集的拉曼光谱的基线校正通过光谱仪自带软件 SLSR Reader V8.3.9 进行。在 MATLAB 软件(美国 MathWorks 公司)中进行小波去噪、归一化、主成分分析和概率神经网络运算。应用 wden 函数实现小波去噪,mapminmax 函数来实现规范化,princomp 函数实现主成分分析,newpnn 函数构造概率神经网络,利用 Minitab 软件(美国 Minitab 股份有限公司)运算得到统计控制图。

---

[1] Hruzikova J, Milde D, Krajancova P, et al. Discrimination of cheese products for authenticity control by infrared spectroscopy[J]. Journal of Agricultural and Food Chemistry, 2012, 60(7): 1845 - 1849.

[2] Zhang Z Y. Rapid discrimination of cheese products based on probabilistic neural network and Raman spectroscopy[J]. Journal of Spectroscopy, 2020, 2020: 8896535.

6.4.2 结果与讨论

1. 乳酪制品的拉曼光谱特征分析

图 6-20 显示了乳酪制品的拉曼光谱，对应的拉曼主要谱峰可能归属如表 6-7 所示。拉曼光谱在 1 760 cm^{-1} 处的峰值主要归因于脂肪酸分子酯的 C=O 伸缩振动。在 1 670 cm^{-1} 处的拉曼峰是蛋白质酰胺 I 的 C=O 伸缩振动和不饱和脂肪酸的 C=C 伸缩振动。1 620 cm^{-1} 处的微弱拉曼光谱可归因于氨基酸苯丙氨酸的苯环振动。光谱的最显著谱峰是脂肪和碳水化合物分子在 1 458 cm^{-1} 处的 CH$_2$ 变形振动。在 1 313 cm^{-1} 处的拉曼峰是与脂质相关的 CH$_2$ 扭曲振动。800～1 200 cm^{-1} 之间的区域主要可归因于碳水化合物，主要包括 C—O 伸缩振动、C—C 伸缩振动和 C—O—H 变形振动（1 143 cm^{-1}，1 095 cm^{-1}，1 080 cm^{-1}）、C—O—C、C—O—H 变形振动和 C—O 伸缩振动（938 cm^{-1}）、C—C—H 和 C—O—C 变形振动（851 cm^{-1}）。这里除了 1 019 cm^{-1} 处的峰值，其是与苯丙氨酸的苯环呼吸振动相关。250～800 cm^{-1} 区域的振动主要包括 C—C—O 变形振动（636 cm^{-1}）、葡萄糖（510 cm^{-1}）和乳糖（384 cm^{-1}）。

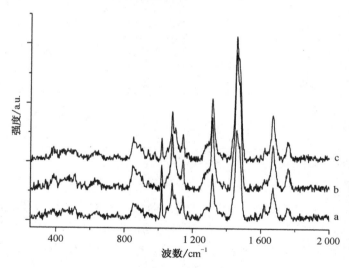

图 6-20 不同品牌 P1(a)、P2(b) 和 P3(c) 乳酪制品的拉曼光谱

不同品牌乳酪制品的拉曼光谱如图 6-20 所示，同时可以看出，不同品牌乳酪制品的拉曼光谱具有很高的相似性，无法通过人眼进行有效的直观判别。同时，乳酪制品的外观是黄色黏稠状固体，也很难从外观上进行样品的识别。通过观察所采集的拉曼光谱图，也发现同一品牌乳酪制品的拉曼光谱强度有一定的波

动,但总体光谱保持较高的一致性。这些不同的品牌样本之间也有很高的相似性,这提示我们需要研究使用统计学习方法进行判别分析。

表6-7 乳酪制品主要的拉曼谱峰及其各自的可能归属

波长/cm^{-1}	归属
1 760	$\nu(C=O)_{酯}$
1 670	$\nu(C=O)$(酰胺Ⅰ),$\nu(C=C)$
1 620	$\nu(C-C)_{环}$
1 458	$\delta(CH_2)$
1 313	$\tau(CH_2)$
1 143	$\nu(C-O)+\nu(C-C)+\delta(C-O-H)$
1 095	$\nu(C-O)+\nu(C-C)+\delta(C-O-H)$
1 080	$\nu(C-O)+\nu(C-C)+\delta(C-O-H)$
1 019	苯丙氨酸苯环振动,$\nu(C-C)_{环}$
938	$\delta(C-O-C)+\delta(C-O-H)+\nu(C-O)$
851	$\delta(C-C-H)+\delta(C-O-C)$
636	$\delta(C-C-O)$
510	葡萄糖
384	乳糖

ν—伸缩振动,δ—变形振动,τ—扭曲振动。

2. 乳酪制品拉曼光谱峰值强度的统计分析

在实际生产管理过程中,经常使用统计过程控制方法对样品质量波动进行统计分析[1]。从以上分析可以看出,拉曼光谱与相应乳酪产品的分子组成密切相关。因此,可以采用统计控制图方法分别分析脂肪(1 760 cm^{-1})、碳水化合物(1 458 cm^{-1})和蛋白质(1 019 cm^{-1})的波动变化情况。

---

[1] Fourie E, Aleixandre-Tudo J L, Mihnea M, et al. Partial least squares calibrations and batch statistical process control to monitor phenolic extraction in red wine fermentations under different maceration conditions[J]. Food Control,2020,115:107303.

统计控制图可以使用以下单值和移动极差控制图公式来实现[①]。对于单值(a)控制图,公式如下:

$$UCL_a = \bar{a} + 2.66\overline{MR}$$

$$CL_a = \bar{a}$$

$$LCL_a = \bar{a} - 2.66\overline{MR}$$

对于移动极差(MR)控制图,公式如下:

$$UCL_{MR} = 3.267\overline{MR}$$

$$CL_{MR} = \overline{MR}$$

$$LCL_{MR} = 0$$

式中,a 和 $\bar{a}$ 分别表示乳酪样品特征峰拉曼强度和平均值;MR 表示移动极差,即 $MR = |a_{i+1} - a_i|$;a_i 代表样品变量 i 处的拉曼强度,在本节工作中 i 以 1 为步长从 1 变为 24。UCL 表示上控制限;LCL 表示下控制限;$\overline{MR}$ 表示移动极差控制图的平均值。

如图 6-21 所示,控制限是基于 P1 品牌在 1 760 cm^{-1} 处的拉曼光谱强度计算的。可以看出,P1 品牌实验样品脂肪含量对应的拉曼峰强度在 55.7~163.1 的范围内波动,移动极差范围位于 0~65.97。实验样品显示出良好的质量稳定性,没有样品跃出控制限。但不难发现,P2 和 P3 品牌的少量实验样本已经跃出了控制限,这表明仅仅监测脂肪含量并不能实现对不同品牌样本的有效区分。同样,图 6-22 和图 6-23 分别显示了根据 P2 和 P3 品牌在 1 760 cm^{-1}(脂肪相关)强度值计算的对照图。图 6-24、图 6-25 和图 6-26 分别显示了根据 P1、P2 和 P3 品牌在 1 458 cm^{-1}(碳水化合物相关)强度值计算的对照图。图 6-27、图 6-28 和图 6-29 分别显示了根据 P1、P2 和 P3 品牌拉曼光谱在 1 019 cm^{-1}(蛋白质相关)强度值计算的对照图。结果表明,基于上述单一指标的控制图可以有效地描述每个品牌实验样本的质量波动情况,但由于样本之间的高度相似性,无法实现品牌有效的差异化分析。

[①] Montgomery D C. Introduction to statistical quality control[M]. 7th ed. Hoboken: John Wiley & Sons,Inc., 2013.

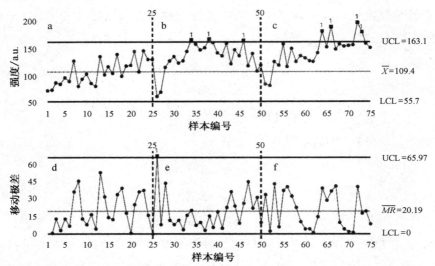

图 6-21 基于在 1 760 cm^{-1} 处拉曼峰值强度的 P1 品牌(a)、P2 品牌(b)和 P3 品牌(c)的质量波动单值控制图,以及 P1 品牌(d)、P2 品牌(e)和 P3 品牌(f)的移动极差控制图

注:UCL 表示上控制限,LCL 表示下控制限,$\overline{X}$ 表示单值控制图的平均值,$\overline{MR}$ 表示移动极差控制图的平均值。基于 P1 品牌的拉曼数据计算得到 UCL、LCL 和 $\overline{MR}$。

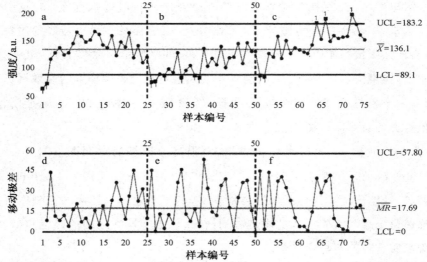

图 6-22 基于在 1 760 cm^{-1} 处拉曼峰值强度的 P2 品牌(a)、P1 品牌(b)和 P3 品牌(c)的质量波动单值控制图,以及 P2 品牌(d)、P1 品牌(e)和 P3 品牌(f)的移动极差控制图

注:UCL 表示上控制限,LCL 表示下控制限,$\overline{X}$ 表示单值控制图的平均值,$\overline{MR}$ 表示移动极差控制图的平均值。基于 P2 品牌的拉曼数据计算得到 UCL、LCL 和 $\overline{MR}$。

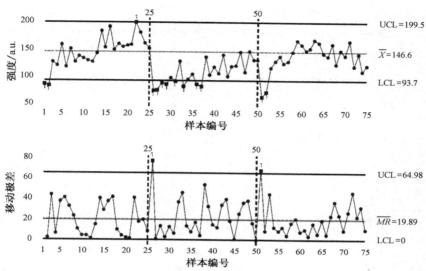

图6-23 基于在 $1\,760\text{ cm}^{-1}$ 处拉曼峰值强度的 P3 品牌(a)、P1 品牌(b)和 P2 品牌(c)的质量波动单值控制图,以及 P3 品牌(d)、P1 品牌(e)和 P2 品牌(f)的移动极差控制图

注:UCL 表示上控制限,LCL 表示下控制限,$\overline{X}$ 表示单值控制图的平均值,$\overline{MR}$ 表示移动极差控制图的平均值。基于 P3 品牌的拉曼数据计算得到 UCL、LCL 和 $\overline{MR}$。

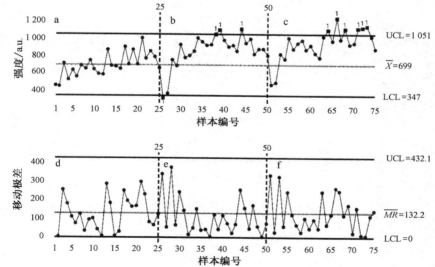

图6-24 基于在 $1\,458\text{ cm}^{-1}$ 处拉曼峰值强度的 P1 品牌(a)、P2 品牌(b)和 P3 品牌(c)的质量波动单值控制图,以及 P1 品牌(d)、P2 品牌(e)和 P3 品牌(f)的移动极差控制图

注:UCL 表示上控制限,LCL 表示下控制限,$\overline{X}$ 表示单值控制图的平均值,$\overline{MR}$ 表示移动极差控制图的平均值。基于 P1 品牌的拉曼数据计算得到 UCL、LCL 和 $\overline{MR}$。

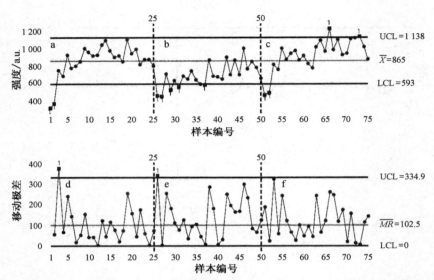

图 6-25 基于在 1 458 cm^{-1} 处拉曼峰值强度的 P2 品牌(a)、P1 品牌(b)和 P3 品牌(c)的质量波动单值控制图,以及 P2 品牌(d)、P1 品牌(e)和 P3 品牌(f)的移动极差控制图

注:UCL 表示上控制限,LCL 表示下控制限,$\overline{X}$ 表示单值控制图的平均值,$\overline{MR}$ 表示移动极差控制图的平均值。基于 P2 品牌的拉曼数据计算得到 UCL、LCL 和 $\overline{MR}$。

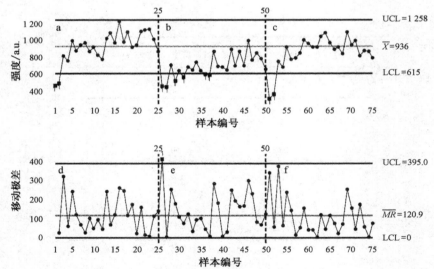

图 6-26 基于在 1 458 cm^{-1} 处拉曼峰值强度的 P3 品牌(a)、P1 品牌(b)和 P2 品牌(c)的质量波动单值控制图,以及 P3 品牌(d)、P1 品牌(e)和 P2 品牌(f)的移动极差控制图

注:UCL 表示上控制限,LCL 表示下控制限,$\overline{X}$ 表示单值控制图的平均值,$\overline{MR}$ 表示移动极差控制图的平均值。基于 P3 品牌的拉曼数据计算得到 UCL、LCL 和 $\overline{MR}$。

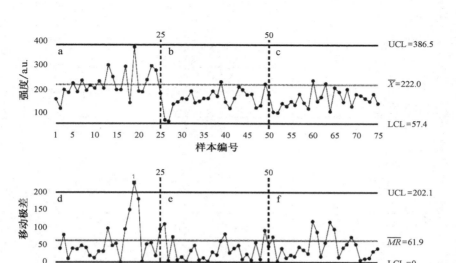

图 6-27 基于在 1 019 cm^{-1} 处拉曼峰值强度的 P1 品牌(a)、P2 品牌(b)和 P3 品牌(c)的质量波动单值控制图,以及 P1 品牌(d)、P2 品牌(e)和 P3 品牌(f)的移动极差控制图

注:UCL 表示上控制限,LCL 表示下控制限,$\overline{X}$ 表示单值控制图的平均值,$\overline{MR}$ 表示移动极差控制图的平均值。基于 P1 品牌的拉曼数据计算得到 UCL、LCL 和 $\overline{MR}$。

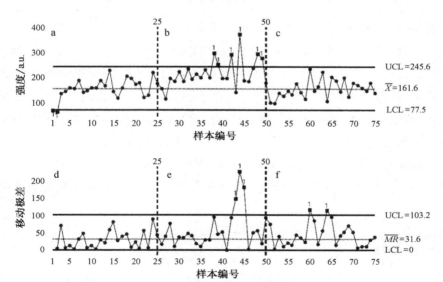

图 6-28 基于在 1 019 cm^{-1} 处拉曼峰值强度的 P2 品牌(a)、P1 品牌(b)和 P3 品牌(c)的质量波动单值控制图,以及 P2 品牌(d)、P1 品牌(e)和 P3 品牌(f)的移动极差控制图

注:UCL 表示上控制限,LCL 表示下控制限,$\overline{X}$ 表示单值控制图的平均值,$\overline{MR}$ 表示移动极差控制图的平均值。基于 P2 品牌的拉曼数据计算得到 UCL、LCL 和 $\overline{MR}$。

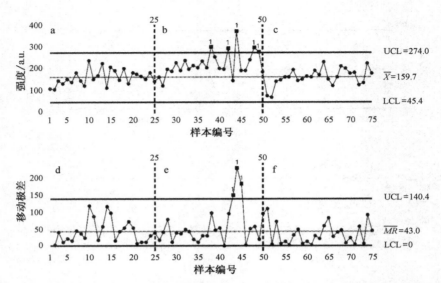

图6-29 基于在 1 019 cm^{-1} 处拉曼峰值强度的 P3 品牌(a)、P1 品牌(b)和 P2 品牌(c)的质量波动单值控制图,以及 P3 品牌(d)、P1 品牌(e)和 P2 品牌(f)的移动极差控制图

注:UCL 表示上控制限,LCL 表示下控制限,$\overline{X}$ 表示单值控制图的平均值,$\overline{MR}$ 表示移动极差控制图的平均值。基于 P3 品牌的拉曼数据计算得到 UCL、LCL 和 $\overline{MR}$。

3. 基于概率神经网络的乳酪制品判别分析

对于这类既有质量波动又有相似性的产品,利用机器学习算法建立有效的判别分析流程已成为研究热点[1]。概率神经网络算法作为一种模式分类算法,具有训练简单、收敛速度快的优点,本节将其用于构造判别分析方法[2][3][4]。该算法由输入层、模型层、求和层和输出层组成。输入层的主要功能是接收训练样本的拉曼光谱数据并将数据传输到网络。神经元的数量一般等于样本的属性维度。模型层主要描述从前一层传递的特征向量以及所有训练样本中每个模式的配对关系。当接收到向量 x 时,该层中类 i 样本的第 j 个神经元的输入输出关系如下

---

[1] Srivastava S, Mishra G, Mishra H N. Probabilistic artificial neural network and E-nose based classification of *Rhyzopertha dominica* infestation in stored rice grains[J]. Chemometrics and Intelligent Laboratory Systems, 2019, 186: 12-22.

[2] 张阳阳, 贾云献, 吴巍屹, 等. 概率神经网络在车辆齿轮箱典型故障诊断中的应用[J]. 汽车工程, 2020, 42(7): 972-977.

[3] Tuccitto N, Bombace A, Torrisi A, et al. Probabilistic neural network-based classifier of ToF-SIMS single-pixel spectra[J]. Chemometrics and Intelligent Laboratory Systems, 2019, 191: 138-142.

[4] Tsuji T, Nobukawa T, Mito A, et al. Recurrent probabilistic neural network-based short-term prediction for acute hypotension and ventricular fibrillation[J]. Scientific Reports, 2020, 10: 11970.

所示：

$$\varphi_{ij}=\frac{1}{(2\pi)^{d/2}\sigma^d}\exp\left[-\frac{(x-x_{ij})^{\mathrm{T}}(x-x_{ij})}{\sigma^2}\right]$$

其中，$i=1,2,\cdots,b$，b 是与训练样本相对应的类的总数。d 是采样空间维度。x_{ij} 是类 i 样本 j 的中心。σ 是平滑因子，并且 φ_{ij} 是模型层中类 i 样本的神经元 j 的输出。

在求和层中，使用以下公式对模型层中属于同一类神经元的输出进行加权和平均：$f_i=\frac{1}{L}\sum_{j=1}^{L}\varphi_{ij}$，这里 f_i 是类 i 的输出，L 是 i 类神经元的数量。

输出层由竞争神经元组成，其功能是接收求和层的输出，并在所有输出层神经元中找到一个后验概率密度最大的神经元。它的输出是预测类别，而其他神经元的输出是 0。公式如下：$y=\mathrm{argmax}(f_i)$，其中 y 是输出层的结果。

可以利用 MATLAB 软件的 newpnn 函数来构建概率神经网络。首先，选择 80% 的实验样本来构建训练集，剩下的 20% 用于构建测试集。实验结果表明，识别准确率仅为 33.33%，原因可能是乳酪样品的拉曼光谱数据中存在冗余信息，影响了模型的有效计算和判别。

根据乳酪制品的拉曼光谱数据，采用小波去噪（wden 函数，db1 小波基，分解层数 = 3）有效地消除了拉曼光谱的谱线噪声，并采用归一化函数（mapminmax 函数）将拉曼光谱强度归属到[−1,1]范围内，从而有效地减少了量纲差异的影响。对于小波去噪，拉曼光谱数据被表示为小波函数的线性组合，即 $f(t)=\sum_{m=-\infty}^{+\infty}\sum_{n=-\infty}^{+\infty}wf(m,n)\psi_{m,n}(t)$。其中 $wf(m,n)$ 是原始拉曼光谱数据 $f(t)$ 中由小波函数 $\psi_{m,n}(t)$ 表示的分量。在计算过程中，将原始拉曼光谱数据转换为小波系数，然后根据软阈值处理方法对较小的系数进行削弱，最后重构去噪后的光谱数据。归一化公式如下：$W=\frac{(W_{\max}-W_{\min})(X-X_{\min})}{X_{\max}-X_{\min}}+W_{\min}$，式中，去噪拉曼光谱数据 X 被映射到 W 的[−1,1]范围内，这里，$W_{\max}=1$ 以及 $W_{\min}=-1$。

如图 6-30 所示，可以看出乳酪制品的拉曼光谱线变得平滑，峰比值有所改善。主成分分析用于提取特征和降低数据维数[①]，在该方法中，一些新的特征变量是通过对原始光谱数据的转换，将原始特征变量线性组合而来的。同时，这些变量应尽可能地表示原始变量的数据特征，并且新的特征变量彼此不相关。基于主成

---

① Almeida M R, de Souza L P, Cesar R S, et al. Investigation of sport supplements quality by Raman spectroscopy and principal component analysis[J]. Vibrational Spectroscopy, 2016, 87: 1-7.

分分析的乳酪制品拉曼光谱特征提取步骤描述如下：(1) 计算拉曼光谱数据矩阵 W 的协方差矩阵 S；(2) 计算协方差矩阵 S 的特征值 $\lambda_1, \lambda_2, \cdots, \lambda_m$（按降序）及其对应的特征向量 $e_1, e_2, \cdots, e_m$；(3) 选择累积贡献率达到一定阈值并形成投影矩阵 $W_{PCA} = [e_1^T, e_2^T, \cdots, e_p^T]$ 的前 p 个特征值对应的特征向量，其中特征值 λ_i 的贡献率定义为 $\dfrac{\lambda_i}{\sum_{k=1}^{m} \lambda_k} \times 100\%$，前 p 个本征值的累积贡献率定义为由 $\dfrac{\sum_{i=1}^{p} \lambda_i}{\sum_{k=1}^{m} \lambda_k} \times 100\%$ 得出；(4) 数据被投影到由特征向量 $Q = W \cdot W_{PCA}$ 形成的空间中，并且 Q 是提取的新特征向量。结果表明，第一主成分可以表征原始1751维数据信息的30.20%，第二主成分可以表征原始信息的11.51%（图6-31）。只需74个主成分即可表征100%的原始信息。图6-32显示了通过主成分分析的乳酪制品的三维散点图，可以看出，同一品牌的实验样本倾向于聚集，不同品牌的实验样本呈现出一些分离和交叉。

随机选取80%的样本作为训练集，将通过小波去噪、归一化和主成分分析提取的74个主成分结果作为数据输入，以剩余20%的样本作为测试集重构概率神经网络模型。实验结果如图6-33所示，最优条件下识别准确率达到100%，平均识别准确率为96%，判别分析耗时不到1 s。实验结果表明，本节工作所建立的拉曼光谱数据采集和处理可以实现对高相似度乳酪制品的有效快速判别，可为其质量控制提供技术支持。

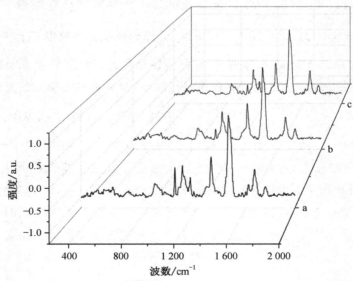

图6-30 不同品牌 P1(a)、P2(b) 和 P3(c) 乳酪制品分别经过小波去噪和归一化后的拉曼光谱图

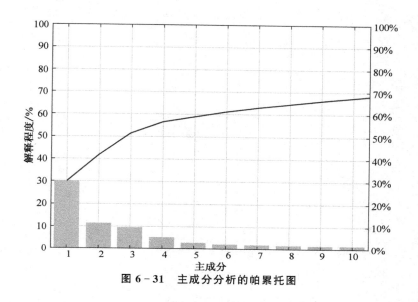

图 6-31　主成分分析的帕累托图

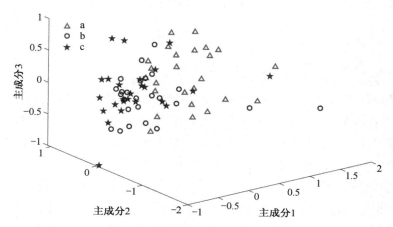

图 6-32　基于拉曼光谱乳酪制品主成分的三维散点图
（a—品牌 P1；b—品牌 P2；c—品牌 P3）

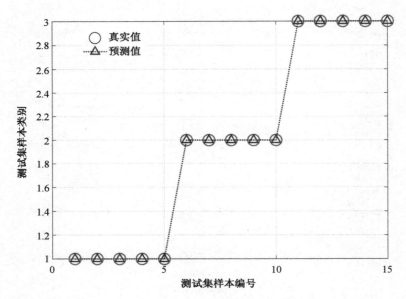

图 6-33 基于概率神经网络算法的乳酪制品实验结果

（y 轴：1—品牌 P1；2—品牌 P2；3—品牌 P3）

6.4.3 小结

本节开发了一种基于拉曼光谱和概率神经网络的新判别分析方法,这将有力地支持解决乳制品质量控制缺乏快速智能判别技术的实际挑战。该方法具有许多优点:乳酪制品的拉曼光谱信号可以直接采集,操作方便快捷,拉曼光谱包含丰富的样品成分和分子振动信息,该方案可以解决传统成分分析和统计控制无法实现的相似品牌有效区分问题。拉曼光谱信号采集、判别算法分析和结果输出的总耗时仅为几分钟,平均识别准确率达到 96%。该方法可供样品间相似度高的食品系统判别分析技术研发参考,具有较高的潜在应用价值。

6.5 拉曼光谱结合模式识别算法的乳制品智能判别与参数优化

乳制品常指以牛奶为原料制备而成的产品,包括奶片、奶粉等,富含多种营养成分,如蛋白质、脂肪、糖类,是居民日常消费品的重要组成部分,其质量安全水平深受国家监管部门和大众的广泛关注。目前,我国已经建立了系列乳制品相关的国家标准,如《食品安全国家标准 乳粉》(GB 19644—2010)规定了牛羊乳粉及调制乳粉的感官要求、理化指标、污染物限量、真菌毒素限量、微生物限量等多项指标以及相应的各类检测方法。针对时有发生的乳制品相关食品质量安全事件,研究人员发展了多种针对乳制品相关非法添加物的测试方法,如杨嘉等运用高效液相色谱法实现了乳粉中三聚氰胺和双氰胺的快速测定[1]。较之常见的色谱法、质谱法,拉曼光谱法是近年来得到快速发展一种散射光谱技术,主要优势在于:样品测试前处理流程少,甚至不需要样品前处理,测试时间短,测试速度快,可实现无损检测等[2]。Nieuwoudt 等使用拉曼光谱结合偏最小二乘法研究了乳制品中三聚氰胺、双氰胺、尿素等的快速判别分析等[3]。

不过,现有检测方法多关注乳制品中的各个指标,如营养成分指标、非法添加物指标等[4],但在品牌类别区分领域的研究报道还相对较为匮乏,因而,迫切需要发展一种快速判别乳制品类别新技术。本节通过拉曼光谱结合模式识别算法用以乳制品判别分析的系统研究,论证了乳制品拉曼光谱数据的采集分析与谱图处理方法,结合识别算法分类器参数优化讨论,提出了适用于乳制品类别智能判别分析的一种新手段[5]。

6.5.1 实验部分

1. 仪器与试剂

实验用乳制品均购置于南京苏果超市,其中,牛初乳奶片标记为 P1,乳酸奶片标记为 P2,两个产品均来自同一家公司生产,风味有所不同,每种乳制品随机选取 40 个样本。

实验仪器采用激光拉曼光谱仪,光谱仪型号:Prott-ezRaman-D3,厂家:美国恩威光电公司(Enwave Optronics)。

---

[1] 杨嘉,丁娟芳,周元元,等.高效液相色谱法快速测定乳粉中的三聚氰胺与双氰胺[J].食品科学,2014,35(6):172-175.

[2] Yang Q Q, Liang F H, Wang D, et al. Simultaneous determination of thiocyanate ion and melamine in milk and milk powder using surface-enhanced Raman spectroscopy[J]. Analytical Methods, 2014, 6(20): 8388-8395.

[3] Nieuwoudt M K, Holroyd S E, McGoverin C M, et al. Raman spectroscopy as an effective screening method for detecting adulteration of milk with small nitrogen-rich molecules and sucrose[J]. Journal of Dairy Science, 2016, 99(4): 2520-2536.

[4] Santos P M, Pereira-Filho E R, Rodriguez-Saona L E. Rapid detection and quantification of milk adulteration using infrared microspectroscopy and chemometrics analysis[J]. Food Chemistry, 2013, 138(1): 19-24.

[5] 王海燕,桂冬冬,沙敏,等.拉曼光谱结合模式识别算法用以牛奶制品智能判别与参数优化[J].中国奶牛,2018(2):55-60.

2. 谱图测试方法

拉曼光谱采集仪器参数设置：激光波长 785 nm，激光功率 450 mW，照射时间 50 s，波长范围 250～2 339 cm^{-1}，分辨率 1 cm^{-1}。乳制品样品直接上样测试。

3. 数据分析

针对两种乳制品的拉曼光谱数据，通过 Kennard-Stone(KS)方法在样本谱图库中选择训练集样本，该方法通过计算样本之间的欧氏距离来选择训练集，能够保证选择的训练集按照空间距离均匀分布[1]。进而建立乳制品拉曼光谱结合支持向量机分类模型，模型选取径向基核函数。径向基核函数是一种非线性函数，可以减少建模过程的计算复杂度，并且提高模型性能[2]。运算平台：MATLAB R2016a。

6.5.2 结果与讨论

1. 乳制品拉曼光谱数据分析

拉曼光谱是一种散射光谱技术，可用以乳制品分子表征，图 6-34 为实验所测牛初乳奶片和乳酸奶片样品库中随机选择的其中某一个样品测试数据绘图所致，参考拉曼光谱用于乳制品测试的相关文献，可对两种乳制品的主要拉曼光谱峰进行归属解析[3]。位于图示最右边的 1 748 cm^{-1} 所示峰可归属于 C=O 伸缩振动，主要可能源自脂肪有关的酯基。1 663 cm^{-1} 所示峰可归属于 C=O 伸缩振动和 C=C 伸缩振动，其中 C=O 伸缩振动可能主要源自蛋白质的酰胺Ⅰ键，C=C 伸缩振动主要源自不饱和脂肪酸。1 465 cm^{-1} 所示峰可归属于 CH_2 变形振动，可能主要源自糖类和脂肪分子。1 337 cm^{-1} 所示峰可能主要归属于糖类的 C—O—H 变形振动、C—O 伸缩振动或游离胆固醇的 C—C 伸缩振动；1 304 cm^{-1} 以及 1 260 cm^{-1} 所示峰可能源自糖类以及饱和脂肪酸的 CH_2 扭曲振动；最高峰 1 130 cm^{-1} 可能主要源自饱和脂肪酸的 C—C 伸缩振动或糖类的 C—C 伸缩振动、C—O 伸缩振动以及 C—O—H 变形振动；1 080 cm^{-1} 所示峰可能主要归属于游离胆固醇的 C—C 伸缩振动或糖类的 C—C 伸缩振动、C—O 伸缩振动以及 C—

---

[1] 李华,王菊香,邢志娜,等. 改进的 K/S 算法对近红外光谱模型传递影响的研究[J]. 光谱学与光谱分析,2011,31(2):362-365.

[2] Wang S J, Liu K S, Yu X J, et al. Application of hybrid image features for fast and non-invasive classification of raisin[J]. Journal of Food Engineering, 2012, 109(3): 531-537.

[3] Almeida M R, de S Oliveira K, Stephani R, et al. Fourier-transform Raman analysis of milk powder: A potential method for rapid quality screening[J]. Journal of Raman Spectroscopy, 2011, 42(7): 1548-1552.

O—H 变形振动；930 cm^{-1} 所示峰可能主要归属于胆固醇或糖类的 C—O—C 变形振动、C—O—H 变形振动和 C—O 伸缩振动；862 cm^{-1} 所示峰主要归属于糖类的 C—C—H 变形振动和 C—O—C 变形振动；800 波数以下还有 9 个峰，依次是 777 cm^{-1}，719 cm^{-1}，652 cm^{-1}，591 cm^{-1}，573 cm^{-1}，518 cm^{-1}，484 cm^{-1}，427 cm^{-1}，363 cm^{-1}，可归属于指纹区，主要可归属于 C—C—O 变形振动、C—S 伸缩振动、C—C—C 变形振动、C—O 扭曲振动等，与乳制品的主要营养成分糖类、脂肪、蛋白质含量密切相关。由图示分析可以看出，拉曼光谱有效反映了乳制品的主要成分信息，可直接用于乳制品结构解析，但是对于具有较高相似性的实验样品的判别分析，仅凭人眼鉴别，存在一定的困难[①]。尤其是随着样品数量的增多，同类别样品内拉曼光谱数据随机波动客观存在，使得人眼判别更加难以实现。因此，运用模式识别分类算法将大大提高拉曼光谱谱图的数据利用效率，实现智能化类别判别分析。

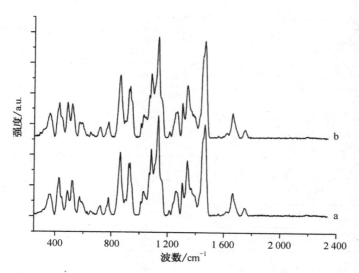

图 6-34　不同品牌 P1(a) 和 P2(b) 乳制品的拉曼光谱谱图

2. 拉曼光谱数据类别判别分析预处理研究

本节实验选取的乳制品均为白色奶片，外观具有极高的相似性，人眼难以判别，采集的拉曼光谱如图 6-34 所示，具有较高的相似性，因此，有必要发展新型判别方法。实验进一步利用随机选取的市售乳制品样品，分别采集其拉曼光谱数

---

① 王海燕，宋超，刘军，等. 基于拉曼光谱—模式识别方法对奶粉进行真伪鉴别和掺伪分析[J]. 光谱学与光谱分析，2017，37(1)：124-128.

据,得到两种乳制品各 40 个样本数据,进行后续模式识别类别判别研究。通过 KS 方法选取出两种乳制品数据各 32 个,得到共 64 个数据组成的训练集,余下的样品数据各 8 个,共 16 个作为测试集。

拉曼光谱测试过程中可能存在高频随机噪音、基线漂移、样品不均匀等情况,进而造成实验数据的变异,通过合理的预处理方法,将有望有效消除拉曼光谱数据中的各种干扰因素,有利于实验建模时有效信息的提取,从而提高模型的准确性和稳定性。实验研究了小波降噪、多元散射校正、一阶导数、二阶导数、归一化以及上述预处理方法间的组合分析[1]。通过不同预处理方法处理后的光谱数据建立支持向量机模型,模型识别率和识别时间结果见表 6-8。

表 6-8 不同预处理方法下的实验结果

预处理方法	识别率(识别样品数/测试集)	识别时间/s
未做预处理情况	50%(8/16)	47.22
WD	50%(8/16)	44.06
MSC	50%(8/16)	42.22
1D	50%(8/16)	40.84
2D	50%(8/16)	40.93
归一化	75%(12/16)	46.70
WD+归一化	93.75%(15/16)	44.95
MSC+归一化	87.5%(14/16)	45.47
1D+归一化	68.75%(11/16)	48.22
2D+归一化	68.75%(11/16)	49.14
WD+MSC+归一化	93.75%(15/16)	45.11
WD+1D+归一化	87.5%(14/16)	48.67
WD+2D+归一化	81.25%(13/16)	49.73
MSC+1D+归一化	87.5%(14/16)	48.71
MSC+2D+归一化	81.25%(13/16)	49.64

---

[1] Zou H Y, Xu K L, Feng Y Y, et al. Application of first order derivative UV spectrophotometry coupled with H-point standard addition to the simultaneous determination of melamine and dicyandiamide in milk[J]. Food Analytical Methods,2015,8(3): 740-748.

续表 6-8

预处理方法	识别率(识别样品数/测试集)	识别时间/s
WD+MSC+1D+归一化	93.75%(15/16)	47.50
WD+MSC+2D+归一化	81.25%(13/16)	49.25

注：WD 表示小波降噪（wavelet denoising），MSC 表示多元散射校正（multiple scattering correction），1D 表示一阶导数（1st derivation），2D 表示二阶导数（2nd derivation），其中，WD 均使用 wden 小波函数（sym5），1D、2D 均使用 Savitzky-Golay 多项式求导，归一化均为[0,1]归一化。

由表 6-8 可以看出，对原始谱图数据进行数据预处理后，可有效提高模型的识别率，同时，不同的数据预处理组合所得的识别率结果存在一定差异。针对建立的支持向量机模型，对比发现：在原始数据没有进行归一化处理的情况下模型最高识别率仅为 50%；而数据经过归一化后的模型识别率为 75%；进一步结合其他预处理方法后，模型最高识别率可达 93.75%。这表明对拉曼数据的归一化等预处理能够显著地提高模型识别率。此外，增加数据预处理步骤，模型的识别时间变化不大。

针对拉曼光谱数据预处理，比较小波降噪、多元散射校正、一阶导数、二阶导数四种预处理方法下拉曼光谱谱图如图 6-35 所示。在拉曼光谱谱图的分析中，一般认为，出现显著峰的信号包含着更多有用的信息，而峰高较小的信号可能就意味着噪声的存在。从图 6-35 可以看出小波降噪和多元散射校正方法能够对拉曼光谱数据进行预处理，除去一定的干扰，并保留谱图的有用信息；而使用一阶导数和二阶导数方法进行谱图数据预处理，拉曼光谱的特征峰信息有效性有所减弱或丢失。同时，对比表 6-8 的四种预处理方法下模型的识别率和识别时间，发现在归一化处理的基础上，小波降噪对于拉曼光谱数据的预处理效果最好；一阶导数和二阶导数对于拉曼光谱数据的预处理在一定程度上会遗失拉曼光谱的特征信号信息，对模型识别率的提高作用不明显。

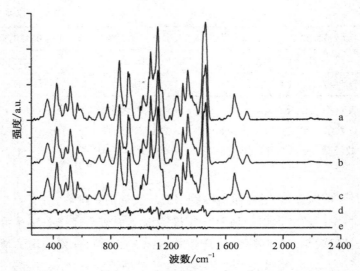

图 6-35 原始光谱(a)、小波降噪(b)、多元散射校正(c)、一阶导数(d)、二阶导数(e)预处理方法下的拉曼光谱谱图

3. 拉曼光谱数据类别判别分析特征提取研究

本次实验中所采集的拉曼光谱数据维度在 2 000 维以上,因而在数据处理上常常可进行数据特征提取。主成分分析是一种常用的数学变换型特征提取方法[①②],通过对实验所得拉曼光谱数据进行主成分分析处理,选取累计贡献率大于 95% 的主成分作为支持向量机模型的输入变量,进一步研究了模型的识别率和识别时间,结果如表 6-9 所示。基于表 6-8 的实验结果,分别选取识别率达到 93.75% 的小波降噪+归一化、小波降噪+多元散射校正+归一化和小波降噪+多元散射校正+一阶导数+归一化这三种预处理方法,与主成分分析结合使用展开后续研究论证。

---

① Fernández Pierna J A, Vincke D, Baeten V, et al. Use of a multivariate moving window PCA for the untargeted detection of contaminants in agro-food products, as exemplified by the detection of melamine levels in milk using vibrational spectroscopy[J]. Chemometrics and Intelligent Laboratory Systems, 2016, 152: 157-162.

② Capuano E, Boerrigter-Eenling R, Koot A, et al. Targeted and untargeted detection of skim milk powder adulteration by near-infrared spectroscopy[J]. Food Analytical Methods, 2015, 8(8): 2125-2134.

表 6-9 主成分分析特征提取下的实验结果

处理方法	识别率(识别样品数/测试集)	识别时间/s
WD+归一化	93.75%(15/16)	44.95
WD+归一化+PCA	93.75%(15/16)	4.22
WD+MSC+归一化	93.75%(15/16)	45.11
WD+MSC+归一化+PCA	93.75%(15/16)	4.10
WD+MSC+1D+归一化	93.75%(15/16)	47.50
WD+MSC+1D+归一化+PCA	100%(16/16)	4.21

注：WD 表示小波降噪（Wavelet Denoising），MSC 表示多元散射校正（Multiple Scattering Correction），PCA 表示主成分分析（Principal Component Analysis）。

通过对比表 6-9 的实验结果，可以看出，数据预处理方法结合主成分分析特征提取后模型识别率保持在 93.75% 甚至可以达到 100%，同时，识别时间从 40~50 s 降至 4~5 s，说明针对拉曼光谱数据进行主成分分析处理，可有效降低冗余信息干扰，提高算法运行效率，减少模型的识别时间消耗。因此，综合考虑模型的识别率和识别时间，可以选取小波降噪+多元散射校正+一阶导数+归一化方法作为拉曼光谱数据的预处理方法，并将预处理后的数据进行主成分分析特征提取，选取累计贡献率大于 95% 的主成分作为支持向量机模型的输入变量。

4. 支持向量机模型核函数参数优化

径向基核函数的惩罚参数 c 和核函数参数 γ 的优化是提高支持向量机模型学习能力与泛化能力的关键[1][2][3]。通过 K-CV（K-fold Cross Validation，K-CV）交叉验证（$k=3$），实验研究了网格搜索算法、粒子群优化算法和遗传算法寻找最佳惩罚参数 c 和核函数参数 γ 值下的模型识别情况[4]，结果见表 6-10。

其中网格搜索算法参数设置为：$c_{min}=-10, c_{max}=5; g_{min}=-10, g_{max}=5$；即参数 c 和 γ 的搜索范围为 $[2^{-10}, 2^5]$，搜索步长为 0.5；粒子群算法参数设置为：惯性

---

[1] Chapelle O, Vapnik V, Bousquet O, et al. Choosing multiple parameters for support vector machines[J]. Machine Learning，2002，46(1)：131-159.

[2] Saini L M, Aggarwal S K, Kumar A. Parameter optimisation using genetic algorithm for support vector machine-based price-forecasting model in National electricity market [J]. IET Generation, Transmission & Distribution，2010，4(1)：36-49.

[3] Domingo E, Tirelli A A, Nunes C A, et al. Melamine detection in milk using vibrational spectroscopy and chemometrics analysis：A review[J]. Food Research International，2014，60：131-139.

[4] Wong T T. Parametric methods for comparing the performance of two classification algorithms evaluated by k-fold cross validation on multiple data sets[J]. Pattern Recognition，2017，65：97-107.

权重 $\omega=1$,速度调节参数 $\eta_1=1.5$,$\eta_2=1.7$,种群数量设为 20,最大进化代数为 100,惩罚参数 c 的搜索范围设置为 $[10^{-1},10^2]$,核函数参数 γ 的搜索范围为 $[10^{-1},10^2]$;遗传算法参数设置为:群体规模为 20,进化代数为 100,交叉概率为 0.9,变异概率为 0.15,参数 c 和 γ 的搜索范围分别为 $[0,100]$ 和 $[0,100]$。表 6-10 结果表明,对比三种寻优算法的模型识别率和算法寻优耗时,采用经网格搜索算法确定的最佳惩罚参数 c 和核函数参数 γ 建立支持向量机模型,具有较高的识别率和较低的寻优耗时。

表 6-10 支持向量机模型参数优化实验结果

算法	网格搜索算法	粒子群优化算法	遗传算法
惩罚参数 c	11.313 7	65.608 5	8.714 3
核函数参数 γ	0.000 976 56	3.667 6	10.418 1
识别率(识别样品数/测试集)	100%(16/16)	50%(8/16)	75%(12/16)
寻优耗时/s	2.72	10.46	13.40

6.5.3 小结

实验选取具有较高相似性的两种乳制品为研究对象,运用拉曼光谱法获取样品的拉曼光谱数据,研究论证了拉曼光谱结合模式识别算法用于乳制品智能判别的可能性。结果显示,实验对象的拉曼光谱蕴含样品丰富的分子振动信息,同时具有高度的相似性,小波降噪、多元散射校正、求导以及归一化相结合的谱图数据预处理,可有效提高拉曼光谱原始数据的信息利用率,有利于算法识别率的提高。主成分分析特征提取算法的使用,可有效降低拉曼光谱数据冗余信息干扰,提高识别算法运行效率,减少模型的识别时间消耗。进一步的参数优化研究,论证了网格搜索算法可得到最佳惩罚参数 $c=11.313\ 7$ 和核函数参数 $\gamma=0.000\ 976\ 56$,据此可建立优化的支持向量机模型。本节的系统研究可为乳制品类别的快速智能化鉴别提供潜在方案,并可为其他食品质量安全快速鉴别方法的发展提供借鉴。

6.6 基于拉曼光谱和极限学习机的乳酪制品表征参数优化

食品作为人们生活的必需品,其质量问题不容忽视。传统的食品质量安全风险主要来自非法添加剂、微生物和重金属污染①②。因此,鉴定和检测方法主要是成分分析,例如用表面增强拉曼光谱法定量测定和分析牛奶中的三聚氰胺、二氰二胺和硫氰酸钠等③④⑤。这种方法可以有效地实现对特定成分的定性和定量分析,但也面临着前置繁琐和耗时的局限性。近年来,出现了一种新的食品质量风险现象。一些廉价的样品被冒充为优质食品⑥,根据国家标准对特定成分的分析,进一步的相关研究发现,其营养成分和污染物限量符合食品质量安全法规的要求,是合格产品。当检测到此类质量问题时,单个成分指标鉴定法很难做出准确的判断⑦。

拉曼光谱作为一种表征分子振动状态的光谱技术,具有直接采集食品样品信号和采样速度快的优点。拉曼光谱与机器学习算法的进一步结合,有望成为解决食品质量判别问题的一个新的研究领域⑧。Huang 等利用拉曼光谱结合化学计量方法对牛奶的酸度和掺假进行了分析⑨,然而,讨论数据处理与算法参数之间的适用关系的文献仍然相对较少,这是乳制品质量判别算法有效判别和应用的关键一步⑩。

---

① Yang L, Wei F, Liu J M, et al. Functional hybrid micro/nanoentities promote agro-food safety inspection[J]. Journal of Agricultural and Food Chemistry, 2021, 69(42): 12402-12417.
② Wang X, Zhao J J, Zhang Q, et al. A chemometric strategy for accurately identifying illegal additive compounds in health foods by using ultra-high-performance liquid chromatography coupled to high resolution mass spectrometry[J]. Analytical Methods, 2021, 13(14): 1731-1739.
③ Hussain A, Pu H B, Sun D W. SERS detection of sodium thiocyanate and benzoic acid preservatives in liquid milk using cysteamine functionalized core-shelled nanoparticles[J]. Spectrochimica Acta Part A, Molecular and Biomolecular Spectroscopy, 2020, 229: 117994.
④ Yang Z J, Zhang R, Chen H, et al. Rapid quantification of thiocyanate in milk samples using a universal paper-based SERS sensor[J]. The Analyst, 2022, 147(22): 5038-5043.
⑤ 吴棉棉,李丹,陆峰. 功能化 SERS 纸基应用于牛奶非法添加物的分离与检测[J]. 光谱学与光谱分析, 2016, 36(S1): 241-242.
⑥ Silva M G, de Paula I L, Stephani R, et al. Raman spectroscopy in the quality analysis of dairy products: A literature review[J]. Journal of Raman Spectroscopy, 2021, 52(12): 2444-2478.
⑦ Lubes G, Goodarzi M. Analysis of volatile compounds by advanced analytical techniques and multivariate chemometrics[J]. Chemical Reviews, 2017, 117(9): 6399-6422.
⑧ Takamura A, Ozawa T. Recent advances of vibrational spectroscopy and chemometrics for forensic biological analysis[J]. The Analyst, 2021, 146(24): 7431-7449.
⑨ Huang W, Fan D S, Li W F, et al. Rapid evaluation of milk acidity and identification of milk adulteration by Raman spectroscopy combined with chemometrics analysis[J]. Vibrational Spectroscopy, 2022, 123: 103440.
⑩ Grelet C, Pierna J A F, Dardenne P, et al. Standardization of milk mid-infrared spectrometers for the transfer and use of multiple models[J]. Journal of Dairy Science, 2017, 100(10): 7910-7921.

在此，本节试图建立一种基于拉曼光谱和极限学习机算法的乳酪制品判别方法，并优化光谱采集的参数。研究工作呈现出三个特点：首先，不同品牌乳酪产品的拉曼光谱相似，将光谱与极限学习机算法相结合，可以在不进行传统成分分析的情况下实现高效判别。其次，实验研究了不同照射时间条件下拉曼光谱数据对判别算法的影响，揭示了合适的光谱信息有利于提高判别算法的准确度，可以为算法的实际推广和使用提供指导。最后，本节建立的判别方法在优化条件下是准确的，具有多种优点，特别是单个样本信号采集仅需 80 s，无需样本预处理，可在线采集，算法运算时间小于 5 s[①]。

6.6.1 实验部分

材料和方法

实验用乳酪制品均购自南京苏果超市，其中，夏洛克乳酪制品标记为品牌 P1，妙可蓝多乳酪标记为品牌 P2，伊利乳酪标记为品牌 P3，每个品牌有 25 个样品。

将适量的样品填充到 96 孔板的每个小孔中，然后使用便携式激光拉曼光谱仪获得每个乳酪样品的拉曼光谱，光谱仪型号：Prott-ezRaman-D3，厂家：美国恩威光电公司（Enwave Optronics）。采集参数包括 785 nm 的激光波长、450 mW 的激光功率、10 s～100 s 的照射时间和 10 s 的间隔、$-85℃$ 的电荷耦合器件温度和 250～2 000 cm^{-1} 的光谱范围，光谱分辨率为 1 cm^{-1}。拉曼光谱仪的激光输出探头具有直径 100 μm 的光斑尺寸。信号采集是在没有任何物理或化学预处理的情况下直接获取的。96 孔板：美国康宁公司（Corning Incorporated）。

通过使用光谱仪自带软件 SLSR Reader V8.3.9 对获得的拉曼光谱进行基线校正。拉曼光谱数据的处理和分析包括小波去噪（wden 函数）、归一化（mapminmax 函数），主成分分析（pca 函数）和极限学习机（elmtrain 和 elmpredict 函数）等，这些都是使用 MATLAB 平台（美国 MathWorks 公司）计算的。

[①] Zhang Z Y, Jiang M Q, Xiong H M. Optimized identification of cheese products based on Raman spectroscopy and an extreme learning machine[J]. New Journal of Chemistry, 2023, 47(14): 6889-6894.

6.6.2 结果与讨论

1. 乳酪制品的拉曼光谱表征和分析

乳酪制品是一种乳黄色的含有水分的固体物质,它的拉曼光谱信号可以被光谱仪直接采集。实验中获得的不同激光照射时间和不同品牌乳酪制品的拉曼光谱如图 6-36 所示。根据现有的相关文献报道,乳酪制品的拉曼光谱可以如表 6-11 所示[1][2]。例如,大约 1 760 cm^{-1} 处的光谱峰主要来自脂肪酸的酯 C=O 伸缩振动,而谱峰最大值出现在大约 1 460 cm^{-1} 处,主要是由于脂肪和糖类的 CH_2 的变形振动。另一个特殊的峰值出现在 1 020 cm^{-1} 左右,主要来自苯丙氨酸的苯环呼吸振动。由于不同品牌乳酪制品的主要成分是脂肪、糖类和蛋白质,它们的拉曼光谱非常相似,每个光谱峰的位置也相似,因此仅凭直观的人眼很难实现相似样品的有效区分。此外,当照射时间为 10 s 时,乳酪制品的拉曼光谱仅具有单独的峰,例如约 1 450 cm^{-1} 和 1 310 cm^{-1}。随着照射时间的增加,光谱峰的数量开始增加,光谱峰值的强度开始显著增加。

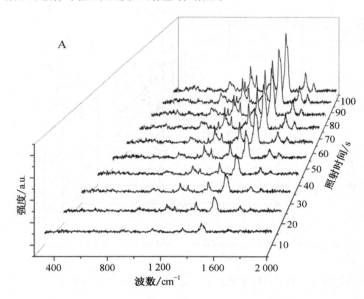

---

[1] Genis D O, Sezer B, Durna S, et al. Determination of milk fat authenticity in ultra-filtered white cheese by using Raman spectroscopy with multivariate data analysis[J]. Food Chemistry, 2021, 336: 127699.

[2] 李靖,李梦银,陈守慧,等. 牛乳主要过敏原的拉曼光谱检测分析[J]. 食品工业,2021,42(12):242-246.

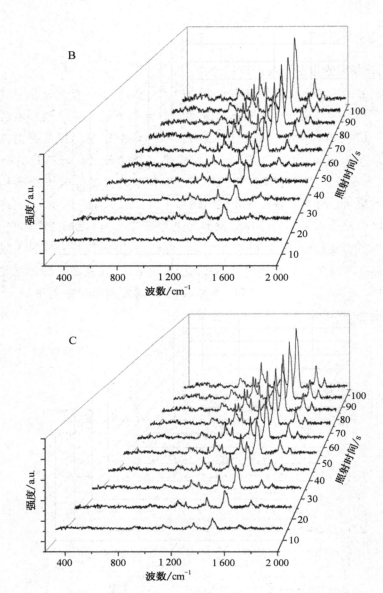

图 6-36　不同品牌 P1(A)、P2(B) 和 P3(C) 乳酪制品在
不同照射时间(10 s~100 s)下的拉曼光谱图

表 6-11 不同品牌乳酪制品的主要拉曼光谱峰的可能归属

P1 波数/cm^{-1}	P2 波数/cm^{-1}	P3 波数/cm^{-1}	归属	可能来源
1 758	1 760	1 758	$\nu(C=O)_{酯}$	脂肪
1 668	1 671	1 668	$\nu(C=O)$酰胺 I	蛋白质
1 668	1 671	1 668	$\nu(C=O),\nu(C=C)$	脂肪
1 570	1 572	1 576	$\delta(N-H),\nu(C-N)$酰胺 II	蛋白质
1 457	1 456	1 456	$\delta(CH_2)$	脂肪
1 457	1 456	1 456	$\delta(CH_2)$	糖类
1 313	1 314	1 314	$\tau(CH_2)$	脂肪
1 143	1 146	1 143	$\nu(C-O)+\nu(C-C)+\delta(C-O-H)$	糖类
1 079	1 080	1 080	$\nu(C-O)+\nu(C-C)+\delta(C-O-H)$	糖类
1 018	1 019	1 019	苯丙氨酸苯环振动,$\nu(C-C)_{环}$	蛋白质
856	853	856	$\delta(C-C-H)+\delta(C-O-C)$	糖类

ν:伸缩振动;δ:变形振动;τ:扭曲振动。

2. 乳酪制品拉曼光谱波动的统计分析

该实验随机收集了来自不同品牌的 25 个样本,并使用欧氏距离测量和质量控制图对其光谱变化进行了统计分析[①]。以品牌 P1 的拉曼光谱数据分析为例,首先计算其光谱的平均值,然后计算每个拉曼光谱与该平均值之间的欧氏距离,最后,将计算结果代入单值和移动极差控制图的运算中,分别得到结果(图 6-37 至图 6-45)。其中,图 6-44 是基于在 80 s 的照射时间获得的拉曼光谱数据计算结果绘制而成。图 6-44(a)和(d)分别显示了品牌 P1 的每个样品的单值和移动极差在平均值附近的波动情况。可以看出,单值图中有两个数据点超过了上控制限,移动极差图中有一个数据点超出了上控制限,其他数据点在控制限内波动,这表明品牌 P1 的各个样本之间存在一定的质量差异,但总体情况仍在可控范围内。图 6-37 至图 6-45 中的各图是根据不同照射时间获得的拉曼光谱数据计算结果绘制得来的,总体上这些结果与图 6-44 所示较为相似。图 6-44(b)和(e)分别是品牌 P2 的单值和移动极差控制图,对应于品牌 P2 的每个样本的拉曼光谱数

---

① Montgomery D C. Introduction to statistical quality control[M]. 7th ed. Hoboken: John Wiley & Sons, Inc., 2013.

据与品牌 P1 的平均值之间的欧氏距离结果,可以看出,只有五个数据点跃出了控制限,表明品牌 P2 的大多数样本与品牌 P1 的样本相似。图 6-44(c)和(f)分别是品牌 P3 的单值和移动极差控制图,对应品牌 P3 的每个样本的拉曼光谱数据与品牌 P1 的平均值之间的欧氏距离结果,结果也与上述相似,只有三个数据点跃出了控制限,表明品牌 P3 的大多数样本与品牌 P1 的样本相似。由上述结果分析不难发现,同一品牌的样品存在质量波动,不同品牌的样品也有很高的相似性,仅使用基于原始拉曼光谱数据与传统的描述性统计分析相结合,距离理想的判别效果尚有一段距离。

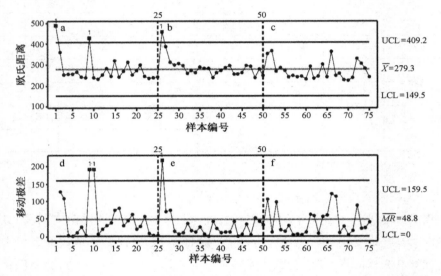

图 6-37 基于拉曼光谱(照射时间:10 s)的欧氏距离计算结果,P1 品牌(a)、P2 品牌(b)和 P3 品牌(c)的质量波动单值以及 P1 品牌(d)、P2 品牌(e)和 P3 品牌(f)的移动极差控制图

注:UCL 表示上控制限,LCL 表示下控制限,$\overline{X}$ 表示单值控制图的平均值,$\overline{MR}$ 表示移动极差控制图的平均值。

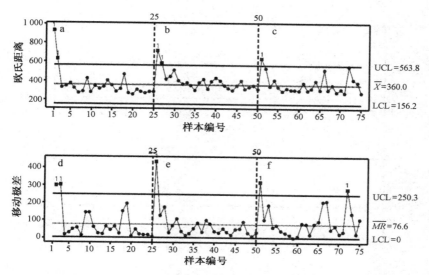

图 6-38 基于拉曼光谱欧氏距离(照射时间:20 s)的计算结果,品牌 P1(a)、品牌 P2(b)和品牌 P3(c)的质量波动单值以及品牌 P1(d)、品牌 P2(e)和品牌 P3(f)的移动极差控制图

注:UCL 表示上控制限,LCL 表示下控制限,$\overline{MR}$表示移动极差控制图的平均值,$\overline{X}$表示单值控制图的平均值。

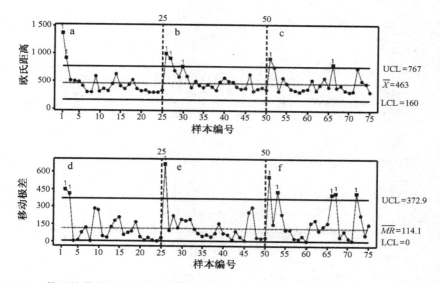

图 6-39 基于拉曼光谱欧氏距离(照射时间:30 s)的计算结果,品牌 P1(a)、品牌 P2(b)和品牌 P3(c)的质量波动单值以及品牌 P1(d)、品牌 P2(e)和品牌 P3(f)的移动极差控制图

注:UCL 表示上控制限,LCL 表示下控制限,$\overline{MR}$表示移动极差控制图的平均值,$\overline{X}$表示单值控制图的平均值。

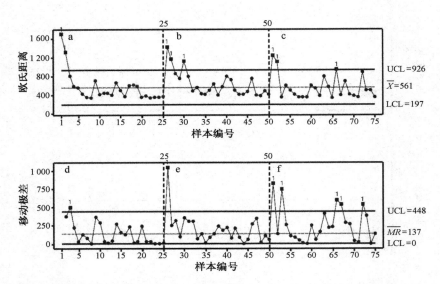

图 6-41 基于拉曼光谱欧氏距离(照射时间:40 s)的计算结果,品牌 P1(a)、品牌 P2(b)和品牌 P3(c)的质量波动单值以及品牌 P1(d)、品牌 P2(e)和品牌 P3(f)的移动极差控制图

注:UCL 表示上控制限,LCL 表示下控制限,$\overline{MR}$ 表示移动极差控制图的平均值,$\overline{X}$ 表示单值控制图的平均值。

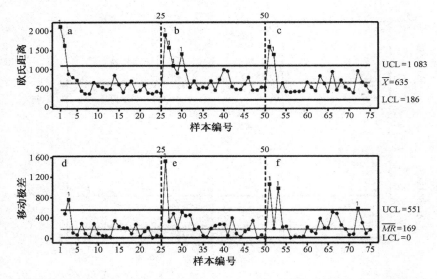

图 6-41 基于拉曼光谱欧氏距离(照射时间:50 s)的计算结果,品牌 P1(a)、品牌 P2(b)和品牌 P3(c)的质量波动单值以及品牌 P1(d)、品牌 P2(e)和品牌 P3(f)的移动极差控制图

注:UCL 表示上控制限,LCL 表示下控制限,$\overline{MR}$ 表示移动极差控制图的平均值,$\overline{X}$ 表示单值控制图的平均值。

第 6 章 基于智能判别算法的乳制品质量智能判别技术研究

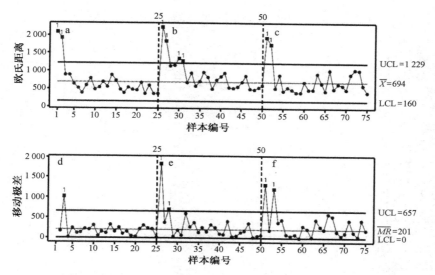

图 6-42 基于拉曼光谱欧氏距离(照射时间:60 s)的计算结果,品牌 P1(a)、品牌 P2(b)和品牌 P3(c)的质量波动单值以及品牌 P1(d)、品牌 P2(e)和品牌 P3(f)的移动极差控制图

注:UCL 表示上控制限,LCL 表示下控制限,$\overline{MR}$ 表示移动极差控制图的平均值,$\overline{X}$ 表示单值控制图的平均值。

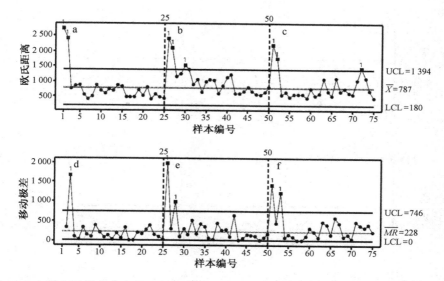

图 6-43 基于拉曼光谱欧氏距离(照射时间:70 s)的计算结果,品牌 P1(a)、品牌 P2(b)和品牌 P3(c)的质量波动单值以及品牌 P1(d)、品牌 P2(e)和品牌 P3(f)的移动极差控制图

注:UCL 表示上控制限,LCL 表示下控制限,$\overline{MR}$ 表示移动极差控制图的平均值,$\overline{X}$ 表示单值控制图的平均值。

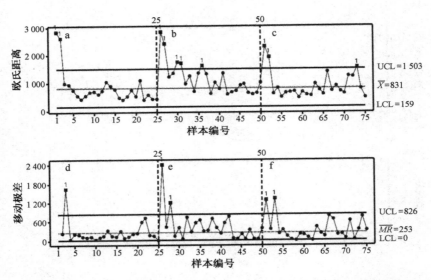

图 6-44 基于拉曼光谱欧氏距离(照射时间:80 s)的计算结果,品牌 P1(a)、品牌 P2(b)和品牌 P3(c)的质量波动单值以及品牌 P1(d)、品牌 P2(e)和品牌 P3(f)的移动极差控制图

注:UCL 表示上控制限,LCL 表示下控制限,$\overline{MR}$ 表示移动极差控制图的平均值,$\overline{X}$ 表示单值控制图的平均值。

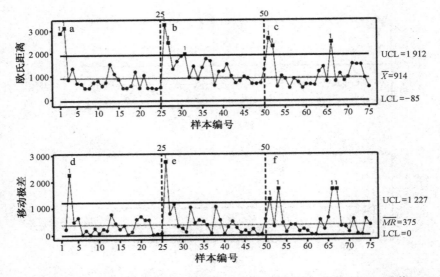

图 6-45 基于拉曼光谱欧氏距离(照射时间:90 s)的计算结果,品牌 P1(a)、品牌 P2(b)和品牌 P3(c)的质量波动单值以及品牌 P1(d)、品牌 P2(e)和品牌 P3(f)的移动极差控制图

注:UCL 表示上控制限,LCL 表示下控制限,$\overline{MR}$表示移动极差控制图的平均值,$\overline{X}$ 表示单值控制图的平均值。

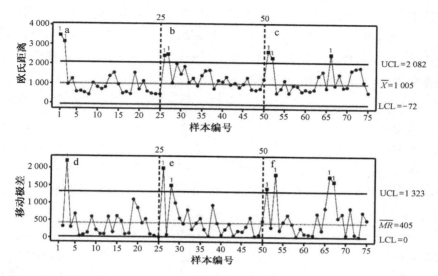

图6-46 基于拉曼光谱欧氏距离(照射时间:100 s)的计算结果,品牌 P1(a)、品牌 P2(b)和品牌 P3(c)的质量波动单值以及品牌 P1(d)、品牌 P2(e)和品牌 P3(f)的移动极差控制图

注:UCL 表示上控制限,LCL 表示下控制限,$\overline{MR}$表示移动极差控制图的平均值,$\overline{X}$ 表示单值控制图的平均值。

3. 乳酪制品拉曼光谱的预处理分析

为了实现对上述相似样本的分类和判别,本节研究工作引入了极限学习机作为一种判别算法,该方法具有学习速度快、泛化性能好等优点[1][2][3]。以照射时间为 80 s 采集的各品牌样本的拉曼光谱为例,将原始数据首次引入极限学习机算法,其中 72% 的样本为测试集,28% 的样本为训练集,样本选择是随机的,运行了 200 次,极限学习机的隐藏层数设置为 250,测试集和训练集的选择均是随机的,结果显示,识别准确率仅为 45%,表明未经优化并不能达到理想的判别效果。由于在采集的拉曼光谱信号中存在一定量的随机噪声,这对判别算法的识别效果存在可能的负面影响,为此,采用小波软阈值去噪方法对这些噪声进行了滤除。实

---

[1] Xiouras C, Cameli F, Quilló G L, et al. Applications of artificial intelligence and machine learning algorithms to crystallization[J]. Chemical Reviews, 2022, 122(15): 13006-13042.

[2] Wu S J, Cui T C, Li Z, et al. Real-time monitoring of the column chromatographic process of Phellodendri Chinensis Cortex part I: End-point determination based on near-infrared spectroscopy combined with machine learning[J]. New Journal of Chemistry, 2022, 46(19): 9085-9097.

[3] Xiao D, Li H Z, Sun X Y. Coal classification method based on improved local receptive field-based extreme learning machine algorithm and visible-infrared spectroscopy[J]. ACS Omega, 2020, 5(40): 25772-25783.

验中使用了coif1小波基,分解层数为5层①。结果如图6-47所示,可以清楚地看出,小波去噪后拉曼光谱变得更平滑。在相同的测试条件下,小波去噪后的识别准确率为62%,较前述未降噪处理的识别结果有所上升。在光谱采集过程中,光谱强度值有一定的波动范围。为了消除量纲的影响,使用归一化方法将光谱强度值归一化为[-1,1](图6-48)。在相同的测试条件下,识别准确率提高到67%。为了进一步提高计算效率,通过主成分分析降低了拉曼光谱数据的维数②。原始拉曼光谱中的每个样本包含1 751个数据点,主成分分析降维后,只需74个提取的主成分即可以代表原始数据中100%的信息。前10个主成分的累计解释帕累托图如图6-49所示,高达77%,其中,第一主成分可以代表40%的原始信息,第二主成分可以代表11%的原始信息;第三主成分可以代表9%的原始信息。如图6-50所示,从图中可以看出,同一品牌的样本有着聚集的趋势,但不同品牌的样本在三维空间中仍显示出一定的交集。在相同的测试条件下,经过主成分分析处理后,识别准确率大大提高到98%,结果表明,主成分降维对极限学习机分类算法的影响是巨大的。

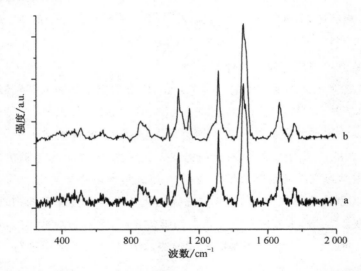

图6-47 P1品牌乳制品的原始拉曼光谱(a)和小波去噪后的光谱(b)(照射时间:80 s)

---

① Xu Y, Zhong P, Jiang A M, et al. Raman spectroscopy coupled with chemometrics for food authentication: A review[J]. TrAC Trends in Analytical Chemistry, 2020, 131: 116017.

② Mozhaeva V, Kudryavtsev D, Prokhorov K, et al. Toxins' classification through Raman spectroscopy with principal component analysis[J]. Spectrochimica Acta Part A, Molecular and Biomolecular Spectroscopy, 2022, 278: 121276.

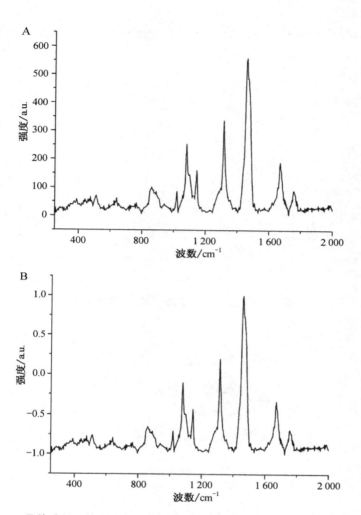

图6-48 P1品牌乳制品的小波去噪后拉曼光谱(A)和归一化光谱(B)(照射时间:80 s)

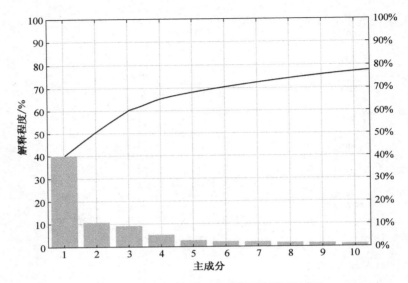

图6-49 主成分分析降维结果的帕累托图

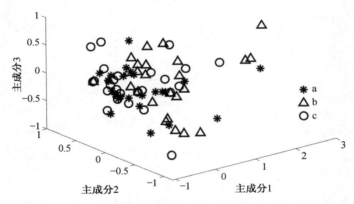

图6-50 不同品牌P1(a)、P2(b)和P3(c)乳酪制品的主成分分析三维散点图

4. 拉曼光谱识别数据的进一步优化分析

进一步研究了不同照射时间乳酪制品的拉曼光谱对判别算法运算结果的影响(图6-51),测试条件是随机选择72%样本作为测试集,其余样本作为训练集,并进行200次随机操作,极限学习机的隐藏层数量为400。图6-51a显示了基于原始拉曼光谱数据的识别结果,可以看出,不同的照射时间对应于不同的识别准确率。随着照射时间的增加,识别准确率呈上升趋势,照射时间为10 s时,最低值为40%,照射时间为80 s时,最高值达到61%,表明在这种条件下很难实现有效判别。小波去噪后,在这种情况下,平均识别准确率变化很小,甚至在50 s的

照射时间内呈下降趋势(图 6-51b)。这种现象与图 6-52 中的情况不同,图 6-52 中的极限学习机的隐藏层数量为 250,经过 40 s 的照射后,小波去噪后的平均识别准确率呈明显上升趋势,这一结果表明,小波去噪的效果与极限学习机的隐藏层数目密切相关。归一化处理结果如图 6-51c 所示,平均识别准确率随着照射时间的增加呈上升趋势,除 80 s 后以外,与图 6-51a 中的识别结果接近。图 6-51d 显示了主成分分析的处理结果,平均识别率呈良好的上升趋势,在 100 s 内最大值达到 85%。图 6-51e 和图 6-51f 分别显示了通过小波去噪和归一化处理、归一化和主成分分析分别结合处理情况下获得的平均识别准确率,可以清楚地看到,组合处理后,平均识别准确率显著提高。图 6-51e 中的最大值为 81%,对应于 100 s 的照射时间;图 6-51f 中的最大值为 97%,对应于 80 s 的照射时间。当将小波去噪、归一化和主成分分析相结合时,结果如图 6-51g 所示,经过 80 s 的照射后,平均识别准确率达到 98%。这一结果表明,与单一处理方法相比,组合预处理技术具有共同提高平均识别准确率的效果。

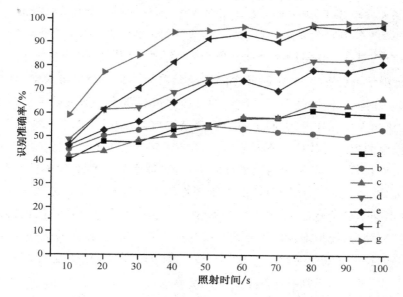

图 6-51 基于原始拉曼光谱(a),小波去噪(b),归一化(c),主成分分析(d),小波去噪与归一化相结合(e),归一化与主成分分析相结合(f)以及小波去噪、归一化与主成分分析相结合(g)的判别结果(隐藏层神经元的数量:400)

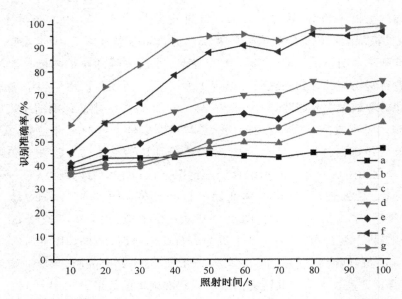

图6-52 基于原始拉曼光谱(a),小波去噪(b),归一化(c),主成分分析(d),小波去噪与归一化相结合(e),归一化与主成分分析相结合(f)以及小波去噪、归一化与主成分分析相结合(g)的判别结果(隐藏层神经元的数量:250)

在极限学习机算法中,隐藏层中的神经元数量也是一个重要因素,有必要对不同的整合时间和隐藏层中不同数量的神经元进行研究。谱图处理包括小波去噪、归一化和主成分分析,它们也用于处理谱图数据,结果如图6-53所示,显示不同隐藏层的神经元数量在不同的照射时间确有一定的差异。在神经元数量处于50~200区间内,随着神经元数量的增加,整体平均识别准确率呈上升趋势。当神经元数量达到250时,识别准确率缓慢增加并趋于稳定。当数量达到400时,识别准确率趋于稳定,基本上变化不大。图6-54显示了基于优化条件的乳酪制品判别结果,显示出良好的分类效果。基于上述研究,在实验优化条件下,即乳酪制品的拉曼光谱照射时间为80s,小波去噪、归一化处理、主成分降维,隐层神经元数量为400个,识别算法的平均识别准确率可达98%,该算法的运算时间也小于5 s。

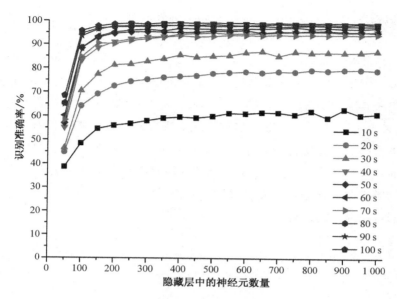

图6-53 隐层神经元数量对极限学习机性能的影响分析

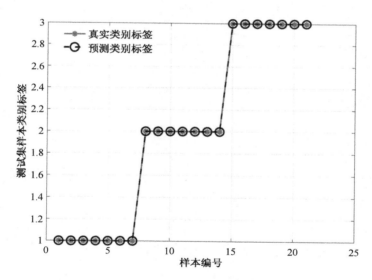

图6-54 基于极限学习机的乳酪制品判别结果

(y轴:标签1代表品牌P1,标签2代表品牌P2,标签3代表品牌P3)

6.6.3 小结

对于相似度较高的乳酪制品,可以直接采集实验样品的拉曼光谱,但光谱也非常相似。仅使用传统的统计分析,结果显示出样本具有一定的质量波动,不过距离理想的判别效果需要再探索。研究进一步开展基于极限学习机算法的判别分析,结果表明,拉曼光谱的照射时间和预处理方法对极限学习机算法的判别效果有很大影响。经过优化,在 80 s 照射时间、小波去噪、归一化、主成分降维、隐层神经元为 400 的条件下,平均识别准确率可达 98%,判别算法运算时间小于 5 s。该方案具有信号采集方便、计算速度快的优点,为有效区分相似样本提供了一条新的可参考研究路径。

参考文献

[1] 钮伟民. 乳及乳制品检测新技术[M]. 北京：化学工业出版社，2012.

[2] Bodor K, Tamási B, Keresztesi Á, et al. A comparative analysis of the nutritional composition of several dairy products in the Romanian market[J]. Heliyon, 2024, 10(11): e31513.

[3] Kaplan M, Baydemir B, Günar B B, et al. Benefits of A2 milk for sports nutrition, health and performance[J]. Frontiers in Nutrition, 2022, 9: 935344.

[4] Salvo E D, Conte F, Casciaro M, et al. Bioactive natural products in donkey and camel milk: A perspective review[J]. Natural Product Research, 2023, 37(12): 2098–2112.

[5] Ho T M, Zou Z Z, Bansal N. Camel milk: A review of its nutritional value, heat stability, and potential food products[J]. Food Research International, 2022, 153: 110870.

[6] da Cunha T M P, Canella M H M, da Silva Haas I C, et al. A theoretical approach to dairy products from membrane processes[J]. Food Science and Technology, 2022, 42: e12522.

[7] Yang L H, Zhu Y Y, Zhang W L, et al. Recent progress in health effects and biosynthesis of lacto-N-tetraose, the most dominant core structure of human milk oligosaccharide[J]. Critical Reviews in Food Science and Nutrition, 2024, 64(19): 6802–6811.

[8] Meng J W, Zhu Y Y, Wang H, et al. Biosynthesis of human milk oligosaccharides: Enzyme cascade and metabolic engineering approaches[J]. Journal of Agricultural and Food Chemistry, 2023, 71(5): 2234–2243.

[9] Kellman B P, Richelle A, Yang J Y, et al. Elucidating Human Milk Oligosaccharide biosynthetic genes through network-based multi-omics integration[J]. Nature Communications, 2022, 13: 2455.

[10] Nagraik R, Sharma A, Kumar D, et al. Milk adulterant detection: Conventional and biosensor based approaches: A review[J]. Sensing and Bio-Sensing Research, 2021, 33: 100433.

[11] Liu S Y, Wang B, Sui Z W, et al. Faster detection of *Staphylococcus aureus* in milk and milk powder by flow cytometry[J]. Foodborne Pathogens and Disease, 2021, 18(5): 346-353.

[12] Zhu Z Z, Guo W C. Recent developments on rapid detection of main constituents in milk: A review[J]. Critical Reviews in Food Science and Nutrition, 2021, 61(2): 312-324.

[13] Yadav M, Kapoor A, Verma A, et al. Functional significance of different milk constituents in modulating the gut microbiome and infant health[J]. Journal of Agricultural and Food Chemistry, 2022, 70(13): 3929-3947.

[14] Shori A B, Al Zahrani A J. Non-dairy plant-based milk products as alternatives to conventional dairy products for delivering probiotics[J]. Food Science and Technology, 2022, 42: e101321.

[15] 包秋华,马学波,任艳,等.应用拉曼光谱对比分析德式乳杆菌保加利亚亚种 ND02 及其 VBNC 态细胞成分[J].食品科学,2022,43(10):114-118.

[16] Wang L L, Shao X Q, Cheng M, et al. Mechanisms and applications of milk-derived bioactive peptides in Food for Special Medical Purposes[J]. International Journal of Food Science & Technology, 2022, 57(5): 2830-2839.

[17] Ma M J, Wang C F, Sun H Y, et al. Precise and efficient HPLC-UV identification of rice glutelin in infant formulas for special medical purposes[J]. International Journal of Food Science & Technology, 2023, 58(7): 3769-3780.

[18] Huang B F, Zhang J S, Wang M L, et al. Determination of vitamin B12 in milk and dairy products by isotope-dilution liquid chromatography tandem mass spectrometry[J]. Journal of Food Quality, 2022, 2022: 7649228.

[19] Canavari M, Coderoni S. Green marketing strategies in the dairy sector: Consumer-stated preferences for carbon footprint labels[J]. Strategic

Change-Briefings in Entrepreneurial Finance, 2019, 28(4): 233-240.

[20] Gao Z F, Li C G, Bai J F, et al. Chinese consumer quality perception and preference of sustainable milk[J]. China Economic Review, 2020, 59: 100939.

[21] Wanniatie V, Sudarwanto M B, Purnawarman T, et al. Milk quality from organic farm[J]. Wartazoa-Buletin Ilmu Peternakan dan Kesehatan Hewan Indonesia, 2017, 27(3): 125-134.

[22] Samarra I, Masdevall C, Foguet-Romero E, et al. Analysis of oxylipins to differentiate between organic and conventional UHT milks[J]. Food Chemistry, 2021, 343: 128477.

[23] 中华人民共和国卫生部. 食品安全国家标准 生乳: GB 19301—2010[S]. 北京: 中国标准出版社, 2010.

[24] 谢琳, 王晓君, 刘彭, 等. 感官检验法鉴别市售纯牛乳制品热处理方式的探讨[J]. 乳业科学与技术, 2008, 31(4): 179-182, 188.

[25] Zeng H, Han H Y, Huang Y D, et al. Rapid prediction of the aroma type of plain yogurts via electronic nose combined with machine learning approaches[J]. Food Bioscience, 2023, 56: 103269.

[26] Perez-Gonzalez C, Salvo-Comino C, Martin-Pedrosa F, et al. A new data analysis approach for an AgNPs-modified impedimetric bioelectronic tongue for dairy analysis[J]. Food Control, 2024, 156: 110136.

[27] Grassi S, Benedetti S, Casiraghi E, et al. E-sensing systems for shelf life evaluation: A review on applications to fresh food of animal origin[J]. Food Packaging and Shelf Life, 2023, 40: 101221.

[28] Wang H Y, Zhang X Y, Yao Y, et al. Oligosaccharide profiles as potential biomarkers for detecting adulteration of caprine dairy products with bovine dairy products[J]. Food Chemistry, 2024, 443: 138551.

[29] Sun Y, Zhao L X, Cai H Y, et al. Composition and factors influencing community structure of lactic acid bacterial in dairy products from Nyingchi Prefecture of Tibet[J]. Journal of Bioscience and Bioengineering, 2023, 135(1): 44-53.

[30] Shishov A, Nizov E, Bulatov A. Microextraction of melamine from dairy products by thymol-nonanoic acid deep eutectic solvent for high-

performance liquid chromatography-ultraviolet determination[J]. Journal of Food Composition and Analysis, 2023, 116: 105083.

[31] Tian H X, Chen S, Li D, et al. Simultaneous detection for adulterations of maltodextrin, sodium carbonate, and whey in raw milk using Raman spectroscopy and chemometrics[J]. Journal of Dairy Science, 2022, 105(9): 7242－7252.

[32] Duan Y F, Chen Y H, Lei M K, et al. Hybrid silica material as a mixed-mode sorbent for solid-phase extraction of hydrophobic and hydrophilic illegal additives from food samples[J]. Journal of Chromatography A, 2022, 1672: 463049.

[33] Sereshti H, Mohammadi Z, Soltani S, et al. Synthesis of a magnetic micro-eutectogel based on a deep eutectic solvent gel immobilized in calcium alginate: Application for green analysis of melamine in milk and dairy products[J]. Talanta, 2023, 265: 124801.

[34] Nanayakkara D, Prashantha M A B, Fernando T L D, et al. Detection and quantification of dicyandiamide (DCD) adulteration in milk using infrared spectroscopy: A rapid and cost-effective screening approach[J]. Food and Humanity, 2023, 1: 1472－1481.

[35] Kumar V, Kaur I, Arora S, et al. Graphene nanoplatelet/graphitized nanodiamond-based nanocomposite for mediator-free electrochemical sensing of urea[J]. Food Chemistry, 2020, 303: 125375.

[36] Nieuwoudt M K, Holroyd S E, McGoverin C M, et al. Raman spectroscopy as an effective screening method for detecting adulteration of milk with small nitrogen-rich molecules and sucrose[J]. Journal of Dairy Science, 2016, 99(4): 2520－2536.

[37] Dong Y L, Yan N, Li X, et al. Rapid and sensitive determination of hydroxyproline in dairy products using micellar electrokinetic chromatography with laser-induced fluorescence detection[J]. Journal of Chromatography A, 2012, 1233: 156－160.

[38] Shawky E, Nahar L, Nassief S M, et al. Dairy products authentication with biomarkers: A comprehensive critical review[J]. Trends in Food Science & Technology, 2024, 147: 104445.

[39] Huang W, Fan D S, Li W F, et al. Rapid evaluation of milk acidity and identification of milk adulteration by Raman spectroscopy combined with chemometrics analysis[J]. Vibrational Spectroscopy, 2022, 123: 103440.

[40] Feng Z K, Liu D, Gu J Y, et al. Raman spectroscopy and fusion machine learning algorithm: A novel approach to identify dairy fraud[J]. Journal of Food Composition and Analysis, 2024, 129: 106090.

[41] Zhao M, Markiewicz-Keszycka M, Beattie R J, et al. Quantification of calcium in infant formula using laser-induced breakdown spectroscopy (LIBS), Fourier transform mid-infrared (FT-IR) and Raman spectroscopy combined with chemometrics including data fusion[J]. Food Chemistry, 2020, 320: 126639.

[42] 邵帅斌,刘美含,石宇晴,等.基于卷积神经网络的乳粉掺杂物拉曼光谱分类方法[J].食品科学,2022,43(14):296-301.

[43] Huang M Y, Yang R J, Zheng Z Y, et al. Discrimination of adulterated milk using temperature-perturbed two-dimensional infrared correlation spectroscopy and multivariate analysis[J]. Spectrochimica Acta Part A, Molecular and Biomolecular Spectroscopy, 2022, 278: 121342.

[44] Wu H Y, Yang R J, Wei Y, et al. Influence of brands on a discrimination model for adulterated milk based on asynchronous two-dimensional correlation spectroscopy slice spectra[J]. Spectrochimica Acta Part A, Molecular and Biomolecular Spectroscopy, 2022, 271: 120958.

[45] Al-Lafi A G, AL-Naser I. Application of 2D-COS-FTIR spectroscopic analysis to milk powder adulteration: Detection of melamine[J]. Journal of Food Composition and Analysis, 2022, 113: 104720.

[46] 陈浩,汪圣尧.仪器分析[M].4版.北京:科学出版社,2022.

[47] 武汉大学.分析化学:下册[M].5版.北京:高等教育出版社,2007.

[48] Hussain Khan H M, McCarthy U, Esmonde-White K, et al. Potential of Raman spectroscopy for in-line measurement of raw milk composition[J]. Food Control, 2023, 152: 109862.

[49] Zhang Z Y, Liu J, Wang H Y. Microchip-based surface enhanced Raman spectroscopy for the determination of sodium thiocyanate in milk[J].

Analytical Letters, 2015, 48(12): 1930-1940.

[50] Yang Q L, Deng X J, Niu B, et al. Qualitative and semi-quantitative analysis of melamine in liquid milk based on surface-enhanced Raman spectroscopy [J]. Spectrochimica Acta Part A, Molecular and Biomolecular Spectroscopy, 2023, 303: 123143.

[51] Nedeljkovic A, Tomasevic I, Miocinovic J, et al. Feasibility of discrimination of dairy creams and cream-like analogues using Raman spectroscopy and chemometric analysis[J]. Food Chemistry, 2017, 232: 487-492.

[52] 徐建华. 我国乳业将迎来更严监管:专家解读上海假奶粉事件[N]. 中国质量报,2016-04-08(2).

[53] 史若天. 探析公共食品安全事件中政府的舆论引导策略:以2016年上海"假奶粉"事件为例[J]. 新闻研究导刊,2016,7(12):334.

[54] 张正勇,沙敏,刘军,等. 基于高通量拉曼光谱的奶粉鉴别技术研究[J]. 中国乳品工业,2017,45(6):49-51.

[55] 张正勇,沙敏,冯楠,等. 基于统计过程控制的液态奶脱脂工序评价分析[J]. 食品安全导刊,2017(28):66-69.

[56] Nieuwoudt M K, Holroyd S E, McGoverin C M, et al. Rapid, sensitive, and reproducible screening of liquid milk for adulterants using a portable Raman spectrometer and a simple, optimized sample well[J]. Journal of Dairy Science, 2016, 99(10): 7821-7831.

[57] 张正勇,李丽萍,岳彤彤,等. 乳粉拉曼光谱表征数据的标准化与降噪处理研究[J]. 粮食科技与经济,2018,43(6):57-61.

[58] 王海燕,等. 食药质量安全检测技术研究[M]. 北京:科学出版社,2023.

[59] Rodrigues Júnior P H, de Sá Oliveira K, de Almeida C E R, et al. FT-Raman and chemometric tools for rapid determination of quality parameters in milk powder: Classification of samples for the presence of lactose and fraud detection by addition of maltodextrin [J]. Food Chemistry, 2016, 196: 584-588.

[60] Zhang Z Y, Sha M, Wang H Y. Laser perturbation two-dimensional correlation Raman spectroscopy for quality control of bovine colostrum products[J]. Journal of Raman Spectroscopy, 2017, 48(8): 1111-1115.

[61] Almeida M R, de S Oliveira K, Stephani R, et al. Fourier-transform Raman analysis of milk powder: A potential method for rapid quality screening[J]. Journal of Raman Spectroscopy, 2011, 42(7): 1548-1552.

[62] 王海燕,宋超,刘军,等.基于拉曼光谱—模式识别方法对奶粉进行真伪鉴别和掺伪分析[J].光谱学与光谱分析,2017,37(1):124-128.

[63] Grelet C, Pierna J A F, Dardenne P, et al. Standardization of milk mid-infrared spectrometers for the transfer and use of multiple models[J]. Journal of Dairy Science, 2017, 100(10): 7910-7921.

[64] 孙雪杉,杨仁杰,杨延荣,等.不同预处理方法对二维相关谱的影响研究Ⅰ:标准化方法[J].天津农学院学报,2015,22(4):13-16,20.

[65] Ehrentreich F, Sümmchen L. Spike removal and denoising of Raman spectra by wavelet transform methods[J]. Analytical Chemistry, 2001, 73(17): 4364-4373.

[66] Avohou T H, Sacré P Y, Hubert P, et al. Interpretable one-class classification of Raman spectra using prediction bands estimated by wavelet regression[J]. Analytical Chemistry, 2022, 94(10): 4183-4191.

[67] 李思维,孙树垒,张正勇.大学生液态奶消费行为研究:以南京市仙林大学城为例[J].粮食科技与经济,2019,44(6):104-108.

[68] 姜冰,李翠霞.基于宏观数据的乳制品质量安全事件的影响及归因分析[J].农业现代化研究,2016,37(1):64-70.

[69] 郑向华,杨自洁,龚吉军,等.基于旋光法的原料乳中乳糖掺伪鉴别技术研究[J].食品工业科技,2015,36(6):75-77,89.

[70] 易冰清,郭秀秀,颜治,等.乳制品掺假现状与稳定同位素鉴别技术研究进展[J].同位素,2020,33(5):293-303.

[71] 赵超敏,王敏,张润何,等.碳氮稳定同位素鉴别有机奶粉[J].现代食品科技,2018,34(12):211-215.

[72] 宋蓓,宋桂雪,宋薇.羊乳制品中牛乳源成分的鉴别检测技术[J].食品科学技术学报,2016,34(6):69-74.

[73] 钱宇,汪慧超,吴林昊,等.超高效液相色谱检测乳粉复原乳中的氨基酸[J].食品与机械,2016,32(7):56-60.

[74] 王之莹,李婷婷,于文杰,等.一种适用于乳制品基因组 DNA 快速提取方法的研究[J].食品安全质量检测学报,2020,11(1):134-139.

[75] 李富威,张舒亚,曾庆坤,等.乳制品中水牛乳成分的实时荧光 PCR 检测技术[J].农业生物技术学报,2013,21(2):247-252.

[76] 王梓笛,李双妹,尹延东,等.基于支持向量机算法的乳制品分类识别技术研究[J].粮食科技与经济,2020,45(3):104-107.

[77] 黄宝莹,佘之蕴,王文敏,等.近红外光谱技术在乳制品快速检测中的应用研究进展[J].中国酿造,2020,39(7):16-19.

[78] 荣菡,甘露菁.基于近红外光谱的自组织映射神经网络快速鉴别牛乳与掺假乳[J].食品工业,2019,40(8):188-191.

[79] 黄文萍,赵依琳,杨如玲,等.基于小波变换的乳酪制品智能鉴别技术[J].粮食科技与经济,2021,46(1):127-130.

[80] 张正勇,岳彤彤,马杰,等.基于拉曼光谱与 k 最近邻算法的酸奶鉴别[J].分析试验室,2019,38(5):553-557.

[81] Li X Z, Wang X N, Wang L Y, et al. Duplex detection of antibiotics in milk powder using lateral-flow assay based on surface-enhanced Raman spectroscopy[J]. Food Analytical Methods, 2021, 14(1): 165-171.

[82] Alvesda Rocha R, Paiva I M, Anjos V, et al. Quantification of whey in fluid milk using confocal Raman microscopy and artificial neural network[J]. Journal of Dairy Science, 2015, 98(6): 3559-3567.

[83] Mendes T O, Junqueira G M A, Porto B L S, et al. Vibrational spectroscopy for milk fat quantification: Line shape analysis of the Raman and infrared spectra[J]. Journal of Raman Spectroscopy, 2016, 47(6): 692-698.

[84] Wang J P, Xie X F, Feng J S, et al. Rapid detection of *Listeria monocytogenes* in milk using confocal micro-Raman spectroscopy and chemometric analysis[J]. International Journal of Food Microbiology, 2015, 204: 66-74.

[85] Cebi N, Dogan C E, Develioglu A, et al. Detection of l-Cysteine in wheat flour by Raman microspectroscopy combined chemometrics of HCA and PCA[J]. Food Chemistry, 2017, 228: 116-124.

[86] Qi M H, Huang X Y, Zhou Y J, et al. Label-free surface-enhanced Raman scattering strategy for rapid detection of penicilloic acid in milk products[J]. Food Chemistry, 2016, 197(Pt A): 723-729.

[87] Zhang Z Y, Gui D D, Sha M, et al. Raman chemical feature extraction for quality control of dairy products[J]. Journal of Dairy Science, 2019, 102(1): 68-76.

[88] Mazurek S, Szostak R, Czaja T, et al. Analysis of milk by FT-Raman spectroscopy[J]. Talanta, 2015, 138: 285-289.

[89] McGoverin C M, Clark A S S, Holroyd S E, et al. Raman spectroscopic quantification of milk powder constituents[J]. Analytica Chimica Acta, 2010, 673(1): 26-32.

[90] Nunes P P, Almeida M R, Pacheco F G, et al. Detection of carbon nanotubes in bovine raw milk through Fourier transform Raman spectroscopy[J]. Journal of Dairy Science, 2024, 107(5): 2681-2689.

[91] Batesttin C, Ângelo F F, Rocha R A, et al. High resolution Raman spectroscopy of raw and UHT bovine and Goat milk[J]. Measurement: Food, 2022, 6: 100029.

[92] Chen J B, Zhou Q, Noda I, et al. Quantitative classification of two-dimensional correlation spectra[J]. Applied Spectroscopy, 2009, 63(8): 920-925.

[93] 王霁月,陆乃彦,季崟,等.人乳与市售婴儿配方乳粉脂质比较研究[J].食品安全质量检测学报,2020,11(21):7784-7790.

[94] Kaleem A, Azmat M, Sharma A, et al. Melamine detection in liquid milk based on selective porous polymer monolith mediated with gold nanospheres by using surface enhanced Raman scattering[J]. Food Chemistry, 2019, 277: 624-631.

[95] Singh P, Singh M K, Beg Y R, et al. A review on spectroscopic methods for determination of nitrite and nitrate in environmental samples[J]. Talanta, 2019, 191: 364-381.

[96] Xue Z H, Zhang Y X, Yu W C, et al. Recent advances in aflatoxin B1 detection based on nanotechnology and nanomaterials-a review[J]. Analytica Chimica Acta, 2019, 1069: 1-27.

[97] Qu L L, Jia Q, Liu C Y, et al. Thin layer chromatography combined with surface-enhanced Raman spectroscopy for rapid sensing aflatoxins[J]. Journal of Chromatography A, 2018, 1579: 115-120.

[98] Stöckel S, Kirchhoff J, Neugebauer U, et al. The application of Raman spectroscopy for the detection and identification of microorganisms[J]. Journal of Raman Spectroscopy, 2016, 47(1): 89-109.

[99] 韩斯琴高娃, 孙佳, 包琳, 等. 基于SERS技术检测牛奶中环丙沙星的研究[J]. 药物分析杂志, 2018, 38(5): 802-805.

[100] Montgomery D C. Introduction to Statistical Quality Control[M]. 6th ed. Hoboken: John Wiley & Sons, Inc., 2009.

[101] Montgomery D C. Introduction to statistical quality control[M]. 7th ed. Hoboken: John Wiley & Sons, Inc., 2013.

[102] Hu X T, Shi J Y, Shi Y Q, et al. Use of a smartphone for visual detection of melamine in milk based on Au@Carbon quantum dots nanocomposites[J]. Food Chemistry, 2019, 272: 58-65.

[103] Wang X, Esquerre C, Downey G, et al. Assessment of infant formula quality and composition using Vis-NIR, MIR and Raman process analytical technologies[J]. Talanta, 2018, 183: 320-328.

[104] Zhang Y S, Zhang Z Y, Zhao Y J, et al. Adaptive compressed sensing of Raman spectroscopic profiling data for discriminative tasks[J]. Talanta, 2020, 211: 120681.

[105] Karacaglar N N Y, Bulat T, Boyaci I H, et al. Raman spectroscopy coupled with chemometric methods for the discrimination of foreign fats and oils in cream and yogurt[J]. Journal of Food and Drug Analysis, 2019, 27(1): 101-110.

[106] Slimani K, Pirotais Y, Maris P, et al. Liquid chromatography-tandem mass spectrometry method for the analysis of N-(3-aminopropyl)-N-dodecylpropane-1, 3-diamine, a biocidal disinfectant, in dairy products[J]. Food Chemistry, 2018, 262: 168-177.

[107] Xiong S Q, Adhikari B, Chen X D, et al. Determination of ultra-low milk fat content using dual-wavelength ultraviolet spectroscopy[J]. Journal of Dairy Science, 2016, 99(12): 9652-9658.

[108] Tan Z, Lou T T, Huang Z X, et al. Single-drop Raman imaging exposes the trace contaminants in milk[J]. Journal of Agricultural and Food Chemistry, 2017, 65(30): 6274-6281.

[109] Weng S Z, Yuan H C, Zhang X Y, et al. Deep learning networks for the recognition and quantitation of surface-enhanced Raman spectroscopy[J]. The Analyst, 2020, 145(14): 4827-4835.

[110] Zhang Z Y. The statistical fusion identification of dairy products based on extracted Raman spectroscopy[J]. RSC Advances, 2020, 10(50): 29682-29687.

[111] Ren S X, Gao L. Improvement of the prediction ability of multivariate calibration by a method based on the combination of data fusion and least squares support vector machines[J]. Analyst, 2011, 136(6): 1252-1261.

[112] Sun H T, Lv G D, Mo J Q, et al. Application of KPCA combined with SVM in Raman spectral discrimination[J]. Optik, 2019, 184: 214-219.

[113] Bakhtiaridoost S, Habibiyan H, Muhammadnejad S, et al. Raman spectroscopy-based label-free cell identification using wavelet transform and support vector machine[J]. RSC Advances, 2016, 6(55): 50027-50033.

[114] 王筠钠,李妍,李扬,等.光谱学技术在稀奶油乳脂肪研究中的应用[J].光谱学与光谱分析,2019,39(6):1773-1778.

[115] Tian F M, Tan F, Li H. An rapid nondestructive testing method for distinguishing rice producing areas based on Raman spectroscopy and support vector machine[J]. Vibrational Spectroscopy, 2020, 107: 103017.

[116] Chen H Z, Xu L L, Ai W, et al. Kernel functions embedded in support vector machine learning models for rapid water pollution assessment via near-infrared spectroscopy[J]. The Science of the Total Environment, 2020, 714: 136765.

[117] Li W L, Yan X, Pan J C, et al. Rapid analysis of the Tanreqing injection by near-infrared spectroscopy combined with least squares support vector machine and Gaussian process modeling techniques[J]. Spectrochimica Acta Part A, Molecular and Biomolecular Spectroscopy, 2019, 218: 271-280.

[118] Hao N, Ping J C, Wang X, et al. Data fusion of near-infrared and mid-

infrared spectroscopy for rapid origin identification and quality evaluation of Lonicerae japonicae flos[J]. Spectrochimica Acta Part A, Molecular and Biomolecular Spectroscopy, 2024, 320: 124590.

[119] Zhou L, Zhang C, Qiu Z J, et al. Information fusion of emerging non-destructive analytical techniques for food quality authentication: A survey[J]. TrAC Trends in Analytical Chemistry, 2020, 127: 115901.

[120] Sha M, Zhang Z Y, Gui D D, et al. Data Fusion of ion Mobility Spectrometry Combined with Hierarchical Clustering Analysis for the Quality Assessment of Apple Essence[J]. Food Analytical Methods, 2017, 10(10): 3415-3423.

[121] Ranveer S A, Harshitha C G, Dasriya V, et al. Assessment of developed paper strip based sensor with pesticide residues in different dairy environmental samples[J]. Current Research in Food Science, 2022, 6: 100416.

[122] Shan J R, Shi L H, Li Y C, et al. SERS-based immunoassay for amplified detection of food hazards: Recent advances and future trends[J]. Trends in Food Science & Technology, 2023, 140: 104149.

[123] Pan W, Liu W J, Huang X J. Rapid identification of the geographical origin of Baimudan tea using a Multi-AdaBoost model integrated with Raman Spectroscopy[J]. Current Research in Food Science, 2024, 8: 100654.

[124] 苏心悦,马艳莉,翟晨,等.表面增强拉曼光谱技术在液体食品品质安全检测中的研究进展[J].光谱学与光谱分析,2023,43(9):2657-2666.

[125] 李海闽,梁琪,陈卫平,等.牛奶中阿莫西林含量表面增强拉曼光谱检测方法的建立[J].食品与机械,2019,35(2):87-91.

[126] Huang W H, Guo L B, Kou W P, et al. Identification of adulterated milk powder based on convolutional neural network and laser-induced breakdown spectroscopy[J]. Microchemical Journal, 2022, 176: 107190.

[127] Yang Z C, Chen G Q, Ma C Q, et al. Magnetic Fe_3O_4@COF@Ag SERS substrate combined with machine learning algorithms for detection of three quinolone antibiotics: Ciprofloxacin, norfloxacin and

levofloxacin[J]. Talanta, 2023, 263: 124725.

[128] Silva M G, de Paula I L, Stephani R, et al. Raman spectroscopy in the quality analysis of dairy products: A literature review[J]. Journal of Raman Spectroscopy, 2021, 52(12): 2444-2478.

[129] Singh P, Pandey S, Manik S. A comprehensive review of the dairy pasteurization process using machine learning models[J]. Food Control, 2024, 164: 110574.

[130] Cui Y W, Lu W B, Xue J, et al. Machine learning-guided REIMS pattern recognition of non-dairy cream, milk fat cream and whipping cream for fraudulence identification[J]. Food Chemistry, 2023, 429: 136986.

[131] Ji H Z, Pu D D, Yan W J, et al. Recent advances and application of machine learning in food flavor prediction and regulation[J]. Trends in Food Science & Technology, 2023, 138: 738-751.

[132] Xue X, Sun H Y, Yang M J, et al. Advances in the application of artificial intelligence-based spectral data interpretation: A perspective[J]. Analytical Chemistry, 2023, 95(37): 13733-13745.

[133] Martini G, Bracci A, Riches L, et al. Machine learning can guide food security efforts when primary data are not available[J]. Nature Food, 2022, 3: 716-728.

[134] Pu H B, Yu J X, Sun D W, et al. Feature construction methods for processing and analysing spectral images and their applications in food quality inspection[J]. Trends in Food Science & Technology, 2023, 138: 726-737.

[135] Wang K Q, Li Z L, Li J J, et al. Raman spectroscopic techniques for nondestructive analysis of agri-foods: A state-of-the-art review[J]. Trends in Food Science & Technology, 2021, 118: 490-504.

[136] Li J X, Qing C C, Wang X Q, et al. Discriminative feature analysis of dairy products based on machine learning algorithms and Raman spectroscopy[J]. Current Research in Food Science, 2024, 8: 100782.

[137] Ni X F, Jiang Y R, Zhang Y S, et al. Identification of liquid milk adulteration using Raman spectroscopy combined with lactose indexed

screening and support vector machine[J]. International Dairy Journal, 2023, 146: 105751.

[138] Song S, Wang Q Y, Zou X, et al. High-precision prediction of blood glucose concentration utilizing Fourier transform Raman spectroscopy and an ensemble machine learning algorithm[J]. Spectrochimica Acta Part A, Molecular and Biomolecular Spectroscopy, 2023, 303: 123176.

[139] Lu B X, Tian F, Chen C, et al. Identification of Chinese red wine origins based on Raman spectroscopy and deep learning[J]. Spectrochimica Acta Part A, Molecular and Biomolecular Spectroscopy, 2023, 291: 122355.

[140] Rong D, Wang H Y, Ying Y B, et al. Peach variety detection using VIS-NIR spectroscopy and deep learning[J]. Computers and Electronics in Agriculture, 2020, 175: 105553.

[141] Wang J F, Lin T H, Ma S Y, et al. The qualitative and quantitative analysis of industrial paraffin contamination levels in rice using spectral pretreatment combined with machine learning models[J]. Journal of Food Composition and Analysis, 2023, 121: 105430.

[142] Chen Z F, Khaireddin Y, Swan A K. Identifying the charge density and dielectric environment of graphene using Raman spectroscopy and deep learning[J]. Analyst, 2022, 147(9): 1824–1832.

[143] Peris M, Escuder-Gilabert L. Electronic noses and tongues to assess food authenticity and adulteration[J]. Trends in Food Science & Technology, 2016, 58: 40–54.

[144] Yang C J, Ding W, Ma L J, et al. Discrimination and characterization of different intensities of goaty flavor in goat milk by means of an electronic nose[J]. Journal of Dairy Science, 2015, 98(1): 55–67.

[145] Poonia A, Jha A, Sharma R, et al. Detection of adulteration in milk: A review[J]. International Journal of Dairy Technology, 2017, 70(1): 23–42.

[146] Zhu W X, Yang J Z, Wang Z X, et al. Rapid determination of 88 veterinary drug residues in milk using automated TurborFlow online clean-up mode coupled to liquid chromatography-tandem mass spectrometry[J]. Talanta, 2016, 148: 401–411.

[147] Tolmacheva V V, Apyari V V, Furletov A A, et al. Facile synthesis of magnetic hypercrosslinked polystyrene and its application in the magnetic solid-phase extraction of sulfonamides from water and milk samples before their HPLC determination[J]. Talanta, 2016, 152: 203 - 210.

[148] Ullah R, Khan S, Khan A, et al. Infant gender-based differentiation in concentration of milk fats using near infrared Raman spectroscopy[J]. Journal of Raman Spectroscopy, 2017, 48(3): 363 - 367.

[149] Chen Y L, Li X L, Yang M, et al. High sensitive detection of penicillin G residues in milk by surface-enhanced Raman scattering[J]. Talanta, 2017, 167: 236 - 241.

[150] Li X Y, Feng S L, Hu Y X, et al. Rapid detection of melamine in milk using immunological separation and surface enhanced Raman spectroscopy[J]. Journal of Food Science, 2015, 80(6): C1196 - C1201.

[151] Hu Y X, Lu X N. Rapid detection of melamine in tap water and milk using conjugated "one-step" molecularly imprinted polymers-surface enhanced Raman spectroscopic sensor[J]. Journal of Food Science, 2016, 81(5): N1272 - N1280.

[152] Noda I. Recent developments in two-dimensional (2D) correlation spectroscopy[J]. Chinese Chemical Letters, 2015, 26(2): 167 - 172.

[153] Noda I, Ozaki Y. Two-Dimensional Correlation Spectroscopy—Applications in Vibrational and Optical Spectroscopy[M]. Chichester: John Wiley & Sons, Inc., 2004.

[154] Eads C D, Noda I. Generalized correlation NMR spectroscopy[J]. Journal of the American Chemical Society, 2002, 124(6): 1111 - 1118.

[155] Park Y, Noda I, Jung Y M. Novel developments and applications of two-dimensional correlation spectroscopy[J]. Journal of Molecular Structure, 2016, 1124: 11 - 28.

[156] Zhang Y L, Chen J B, Lei Y, et al. Discrimination of different red wine by Fourier-transform infrared and two-dimensional infrared correlation spectroscopy[J]. Journal of Molecular Structure, 2010, 974(1/2/3): 144 - 150.

[157] Lei Y, Zhou Q, Zhang Y L, et al. Analysis of crystallized lactose in milk powder by Fourier-transform infrared spectroscopy combined with two-dimensional correlation infrared spectroscopy [J]. Journal of Molecular Structure, 2010, 974(1/2/3): 88 – 93.

[158] Moros J, Javier Laserna J. Unveiling the identity of distant targets through advanced Raman-laser-induced breakdown spectroscopy data fusion strategies[J]. Talanta, 2015, 134: 627 – 639.

[159] Rygula A, Majzner K, Marzec K M, et al. Raman spectroscopy of proteins: A review[J]. Journal of Raman Spectroscopy, 2013, 44(8): 1061 – 1076.

[160] Moros J, Garrigues S, de la Guardia M. Evaluation of nutritional parameters in infant formulas and powdered milk by Raman spectroscopy[J]. Analytica Chimica Acta, 2007, 593(1): 30 – 38.

[161] Zhou Q, Sun S Q, Yu L, et al. Sequential changes of main components in different kinds of milk powders using two-dimensional infrared correlation analysis[J]. Journal of Molecular Structure, 2006, 799(1/2/3): 77 – 84.

[162] Dahlenborg H, Millqvist-Fureby A, Brandner B D, et al. Study of the porous structure of white chocolate by confocal Raman microscopy[J]. European Journal of Lipid Science and Technology, 2012, 114(8): 919 – 926.

[163] Yang R J, Liu R, Dong G M, et al. Two-dimensional hetero-spectral mid-infrared and near-infrared correlation spectroscopy for discrimination adulterated milk [J]. Spectrochimica Acta Part A, Molecular and Biomolecular Spectroscopy, 2016, 157: 50 – 54.

[164] Chen X W, Ye N S. Graphene oxide-reinforced hollow fiber solid-phase microextraction coupled with high-performance liquid chromatography for the determination of cephalosporins in milk samples[J]. Food Analytical Methods, 2016, 9(9): 2452 – 2462.

[165] Gliszczyńska-Świgło A, Chmielewski J. Electronic nose as a tool for monitoring the authenticity of food. A review[J]. Food Analytical Methods, 2017, 10(6): 1800 – 1816.

[166] Alsammarraie F K, Lin M S. Using standing gold nanorod arrays as surface-enhanced Raman spectroscopy (SERS) substrates for detection of carbaryl residues in fruit juice and milk[J]. Journal of Agricultural and Food Chemistry, 2017, 65(3): 666-674.

[167] Ritota M, Manzi P. Melamine detection in milk and dairy products: Traditional analytical methods and recent developments[J]. Food Analytical Methods, 2018, 11(1): 128-147.

[168] 宁霄,金绍明,李志远,等.功能性老年乳粉中300种非法添加药物及其类似物的液相色谱—高分辨质谱分析[J].色谱,2023,41(11):960-975.

[169] Park Y, Jin S L, Noda I, et al. Recent progresses in two-dimensional correlation spectroscopy (2D-COS)[J]. Journal of Molecular Structure, 2018, 1168: 1-21.

[170] 班晶晶,刘贵珊,何建国,等.基于表面增强拉曼光谱与二维相关光谱法检测鸡肉中恩诺沙星残留[J].食品与机械,2020,36(7):55-58.

[171] Zhang Z Y, Li S W, Sha M, et al. Characterization of fresh milk products based on multidimensional Raman spectroscopy[J]. Journal of Applied Spectroscopy, 2021, 87(6): 1206-1215.

[172] Ahmad N, Saleem M. Raman spectroscopy based characterization of desi ghee obtained from buffalo and cow milk[J]. International Dairy Journal, 2019, 89: 119-128.

[173] Amjad A, Ullah R, Khan S, et al. Raman spectroscopy based analysis of milk using random forest classification[J]. Vibrational Spectroscopy, 2018, 99: 124-129.

[174] Li-Chan E C Y. The applications of Raman spectroscopy in food science[J]. Trends in Food Science & Technology, 1996, 7(11): 361-370.

[175] Hoang V D. Wavelet-based spectral analysis[J]. TrAC Trends in Analytical Chemistry, 2014, 62: 144-153.

[176] He W, Zhou J, Cheng H, et al. Validation of origins of tea samples using partial least squares analysis and Euclidean distance method with near-infrared spectroscopy data[J]. Spectrochimica Acta Part A: Molecular and Biomolecular Spectroscopy, 2012, 86: 399-404.

[177] Chen J B, Zhou Q, Noda I, et al. Discrimination of different Genera

[178] Xie Y, You Q B, Dai P Y, et al. How to achieve auto-identification in Raman analysis by spectral feature extraction & Adaptive Hypergraph[J]. Spectrochimica Acta Part A, Molecular and Biomolecular Spectroscopy, 2019, 222: 117086.

[179] Ahmad N, Saleem M. Characterization of desi ghee obtained from different extraction methods using Raman spectroscopy [J]. Spectrochimica Acta Part A, Molecular and Biomolecular Spectroscopy, 2019, 223: 117311.

[180] 张淑萍,陆娟. 我国乳品行业市场发展整体状况研究[J]. 中国乳品工业, 2013,41(11):33-37.

[181] 刘回春. 郴州"大头娃娃"奶粉事件定性:母婴店虚假宣传[J]. 中国质量万里行,2020(6):68-69.

[182] 中华人民共和国卫生部. 食品安全国家标准 乳粉:GB 19644—2010[S]. 北京:中国标准出版社,2010.

[183] 国家质量监督检验检疫总局. 原料乳与乳制品中三聚氰胺检测方法:GB/T 22388—2008[S]. 北京:中国标准出版社,2008.

[184] 崔向云,常建军,张雪峰,等. UPLC-MS/MS 法测定液态乳中舒巴坦残留[J]. 中国乳品工业,2014,42(10):42-43,45.

[185] 王海燕,张庆民. 质量分析与质量控制[M]. 北京:电子工业出版社,2015.

[186] 刘锐,魏益民,张波. 基于统计过程控制(SPC)的挂面加工过程质量控制[J]. 食品科学,2013,34(8):43-47.

[187] 王立甜. 聚焦"十四五"食品安全[N]. 中国市场监管报,2021-12-07(3).

[188] 李涛,周艳华,向俊,等. 分散固相萃取—超高效液相色谱—三重四极杆质谱法快速检测奶粉中 16 种喹诺酮药物残留[J]. 中国乳品工业,2021, 49(11):54-58,64.

[189] de Oliveira Mendes T, Manzolli Rodrigues B V, Simas Porto B L, et al. Raman Spectroscopy as a fast tool for whey quantification in raw milk[J]. Vibrational Spectroscopy, 2020, 111: 103150.

[190] Li Y, Tang S S, Zhang W J, et al. A surface-enhanced Raman

scattering-based lateral flow immunosensor for colistin in raw milk[J]. Sensors and Actuators B：Chemical，2019，282：703-711.

[191] 顾颖,赵姝彤,熊蓝萍,等.基于多维拉曼光谱的乳粉表征及判别分析[J].粮食科技与经济,2022,47(3):97-101.

[192] dos Santos Pereira E V，de Sousa Fernandes D D，de Araújo M C U，et al. Simultaneous determination of goat milk adulteration with cow milk and their fat and protein contents using NIR spectroscopy and PLS algorithms[J]. LWT，2020，127：109427.

[193] Genis D O，Sezer B，Durna S，et al. Determination of milk fat authenticity in ultra-filtered white cheese by using Raman spectroscopy with multivariate data analysis[J]. Food Chemistry，2021，336：127699.

[194] 罗洁,王紫薇,宋君红,等.不同品种牛乳脂质的共聚焦拉曼光谱指纹图谱[J].光谱学与光谱分析,2016,36(1):125-129.

[195] Marcott C，Kansiz M，Dillon E，et al. Two-dimensional correlation analysis of highly spatially resolved simultaneous IR and Raman spectral imaging of bioplastics composite using optical photothermal Infrared and Raman spectroscopy[J]. Journal of Molecular Structure，2020，1210：128045.

[196] 中华人民共和国卫生部.食品安全国家标准　发酵乳:GB 19302—2010[S].北京:中国标准出版社,2010.

[197] Hua M Z，Feng S L，Wang S，et al. Rapid detection and quantification of 2，4-dichlorophenoxyacetic acid in milk using molecularly imprinted polymers-surface-enhanced Raman spectroscopy[J]. Food Chemistry，2018，258：254-259.

[198] 刘文涵,杨未,张丹.苯丙氨酸银溶胶表面增强拉曼光谱的研究[J].光谱学与光谱分析,2008,28(2):343-346.

[199] 梅江元.基于马氏距离的度量学习算法研究及应用[D].哈尔滨:哈尔滨工业大学,2016.

[200] 剧柠,胡婕.光谱技术在乳及乳制品研究中的应用进展[J].食品与机械,2019,35(1):232-236.

[201] 郭文辉,袁彩霞,洪霞,等.乳制品中氰化物的快速检测[J].中国乳品工

业,2019,47(2):61-64.

[202] 张群.乳制品中抗生素的荧光快速检测技术研究及应用[J].食品与生物技术学报,2018,37(12):1336.

[203] 石彬,李咏富,吴远根.氯化血红素比色法检测乳制品中土霉素[J].中国酿造,2018,37(7):168-172.

[204] 赵小旭,柳家鹏,柴艳兵,等.胶体金免疫层析法快速检测乳制品中重金属离子铅[J].粮食科技与经济,2018,43(3):51-54.

[205] Fang S Y, Wu S Y, Chen Z, et al. Recent progress and applications of Raman spectrum denoising algorithms in chemical and biological analyses: A review[J]. TrAC Trends in Analytical Chemistry, 2024, 172: 117578.

[206] 李志豪,沈俊,边瑞华,等.机器学习算法用于公安一线拉曼实际样本采样学习及其准确度比较[J].光谱学与光谱分析,2019,39(7):2171-2175.

[207] 陈思雨,张舒慧,张纾,等.基于共聚焦拉曼光谱技术的苹果轻微损伤早期判别分析[J].光谱学与光谱分析,2018,38(2):430-435.

[208] Du L J, Lu W Y, Cai Z J, et al. Rapid detection of milk adulteration using intact protein flow injection mass spectrometric fingerprints combined with chemometrics[J]. Food Chemistry, 2018, 240: 573-578.

[209] 张文雅,范雨强,韩华,等.基于交叉验证网格寻优支持向量机的产品销售预测[J].计算机系统应用,2019,28(5):1-9.

[210] Zhang L G, Zhang X, Ni L J, et al. Rapid identification of adulterated cow milk by non-linear pattern recognition methods based on near infrared spectroscopy[J]. Food Chemistry, 2014, 145: 342-348.

[211] Battisti I, Ebinezer L B, Lomolino G, et al. Protein profile of commercial soybean milks analyzed by label-free quantitative proteomics[J]. Food Chemistry, 2021, 352: 129299.

[212] Schulmerich M V, Walsh M J, Gelber M K, et al. Protein and oil composition predictions of single soybeans by transmission Raman spectroscopy[J]. Journal of Agricultural and Food Chemistry, 2012, 60(33): 8097-8102.

[213] Yin H C, Huang J, Zhang H R. Study on isolation and Raman

spectroscopy of glycinin in soybean protein[J]. Grain & Oil Science and Technology, 2018, 1(2): 72-76.

[214] Vasafi P S, Hinrichs J, Hitzmann B. Establishing a novel procedure to detect deviations from standard milk processing by using online Raman spectroscopy[J]. Food Control, 2022, 131: 108442.

[215] Li L M, Chin W S. Rapid and sensitive SERS detection of melamine in milk using Ag nanocube array substrate coupled with multivariate analysis[J]. Food Chemistry, 2021, 357: 129717.

[216] Xi J, Yu Q R. The development of lateral flow immunoassay strip tests based on surface enhanced Raman spectroscopy coupled with gold nanoparticles for the rapid detection of soybean allergen β-conglycinin[J]. Spectrochimica Acta Part A, Molecular and Biomolecular Spectroscopy, 2020, 241: 118640.

[217] Zhao H F, Zhan Y L, Xu Z, et al. The application of machine-learning and Raman spectroscopy for the rapid detection of edible oils type and adulteration[J]. Food Chemistry, 2022, 373(Pt B): 131471.

[218] Barros I H A S, Santos L P, Filgueiras P R, et al. Design experiments to detect and quantify soybean oil in extra virgin olive oil using portable Raman spectroscopy[J]. Vibrational Spectroscopy, 2021, 116: 103294.

[219] Li Y, Zhang J Y, Wang Y Z. FT-MIR and NIR spectral data fusion: A synergetic strategy for the geographical traceability of *Panax notoginseng*[J]. Analytical and Bioanalytical Chemistry, 2018, 410(1): 91-103.

[220] Zhang Z Y, Shi X J, Zhao Y J, et al. Brand Identification of Soybean Milk Powder based on Raman Spectroscopy Combined with Random Forest Algorithm[J]. Journal of Analytical Chemistry, 2022, 77(10): 1282-1286.

[221] Wu L K, Wang L M, Qi B K, et al. 3D confocal Raman imaging of oil-rich emulsion from enzyme-assisted aqueous extraction of extruded soybean powder[J]. Food Chemistry, 2018, 249: 16-21.

[222] El-Abassy R M, Eravuchira P J, Donfack P, et al. Fast determination of milk fat content using Raman spectroscopy[J]. Vibrational

Spectroscopy, 2011, 56(1): 3-8.

[223] Lee H, Cho B K, Kim M S, et al. Prediction of crude protein and oil content of soybeans using Raman spectroscopy[J]. Sensors and Actuators B: Chemical, 2013, 185: 694-700.

[224] Jin H Q, Li H, Yin Z K, et al. Application of Raman spectroscopy in the rapid detection of waste cooking oil[J]. Food Chemistry, 2021, 362: 130191.

[225] Sha M, Gui D D, Zhang Z Y, et al. Evaluation of sample pretreatment method for geographic authentication of rice using Raman spectroscopy[J]. Journal of Food Measurement and Characterization, 2019, 13(3): 1705-1712.

[226] Zhao X Y, Liu G Y, Sui Y T, et al. Denoising method for Raman spectra with low signal-to-noise ratio based on feature extraction[J]. Spectrochimica Acta Part A, Molecular and Biomolecular Spectroscopy, 2021, 250: 119374.

[227] Hu J Q, Zhang D, Zhao H T, et al. Intelligent spectral algorithm for pigments visualization, classification and identification based on Raman spectra[J]. Spectrochimica Acta Part A, Molecular and Biomolecular Spectroscopy, 2021, 250: 119390.

[228] 陈凤霞,杨天伟,李杰庆,等.基于偏最小二乘法判别分析与随机森林算法的牛肝菌种类鉴别[J].光谱学与光谱分析,2022,42(2):549-554.

[229] Li M G, Xu Y Y, Men J, et al. Hybrid variable selection strategy coupled with random forest (RF) for quantitative analysis of methanol in methanol-gasoline via Raman spectroscopy[J]. Spectrochimica Acta Part A, Molecular and Biomolecular Spectroscopy, 2021, 251: 119430.

[230] de Santana F B, Mazivila S J, Gontijo L C, et al. Rapid discrimination between authentic and adulterated andiroba oil using FTIR-HATR spectroscopy and random forest[J]. Food Analytical Methods, 2018, 11(7): 1927-1935.

[231] Wang J J, Wu Y, Wu Q H, et al. Highly sensitive detection of melamine in milk samples based on N-methylmesoporphyrin IX/G-quadruplex structure[J]. Microchemical Journal, 2020, 155: 104751.

[232] Zhou W E, Wu H Q, Wang Q, et al. Simultaneous determination of formononetin, biochanin A and their active metabolites in human breast milk, saliva and urine using salting-out assisted liquid-liquid extraction and ultra high performance liquid chromatography-electrospray ionization tandem mass spectrum[J]. Journal of Chromatography B, Analytical Technologies in the Biomedical and Life Sciences, 2020, 1145: 122108.

[233] Ramezani A M, Ahmadi R, Absalan G. Designing a sustainable mobile phase composition for melamine monitoring in milk samples based on micellar liquid chromatography and natural deep eutectic solvent[J]. Journal of Chromatography A, 2020, 1610: 460563.

[234] Shuib N S, Makahleh A, Salhimi S M, et al. Determination of aflatoxin M_1 in milk and dairy products using high performance liquid chromatography-fluorescence with post column photochemical derivatization[J]. Journal of Chromatography A, 2017, 1510: 51-56.

[235] Yan S, Lai X X, Du G R, et al. Identification of aminoglycoside antibiotics in milk matrix with a colorimetric sensor array and pattern recognition methods[J]. Analytica Chimica Acta, 2018, 1034: 153-160.

[236] Na G Q, Hu X F, Yang J F, et al. Colloidal gold-based immunochromatographic strip assay for the rapid detection of bacitracin zinc in milk[J]. Food Chemistry, 2020, 327: 126879.

[237] Luan Q, Gan N, Cao Y T, et al. Mimicking an enzyme-based colorimetric aptasensor for antibiotic residue detection in milk combining magnetic loop-DNA probes and CHA-assisted target recycling amplification[J]. Journal of Agricultural and Food Chemistry, 2017, 65(28): 5731-5740.

[238] Ai K L, Liu Y L, Lu L H. Hydrogen-bonding recognition-induced color change of gold nanoparticles for visual detection of melamine in raw milk and infant formula[J]. Journal of the American Chemical Society, 2009, 131(27): 9496-9497.

[239] Shi Q Q, Huang J, Sun Y N, et al. Utilization of a lateral flow colloidal

gold immunoassay strip based on surface-enhanced Raman spectroscopy for ultrasensitive detection of antibiotics in milk[J]. Spectrochimica Acta Part A，Molecular and Biomolecular Spectroscopy，2018，197：107 - 113.

[240] Shin W R，Sekhon S S，Rhee S K，et al. Aptamer-based paper strip sensor for detecting Vibrio fischeri[J]. ACS Combinatorial Science，2018，20(5)：261 - 268.

[241] Balan B J，Dhaulaniya A S，Jamwal R，et al. Application of Attenuated Total Reflectance-Fourier Transform Infrared （ATR-FTIR）spectroscopy coupled with chemometrics for detection and quantification of formalin in cow milk[J]. Vibrational Spectroscopy，2020，107：103033.

[242] Lu W Y，Liu J，Gao B Y，et al. Technical note：Nontargeted detection of adulterated plant proteins in raw milk by UPLC-quadrupole time-of-flight mass spectrometric proteomics combined with chemometrics[J]. Journal of Dairy Science，2017，100(9)：6980 - 6986.

[243] Teixeira R C，Luiz L C，Junqueira G M A，et al. Detection of antibiotic residues in Cow's milk：A theoretical and experimental vibrational study[J]. Journal of Molecular Structure，2020，1215：128221.

[244] 冯彦婷,林沛纯,谢慧风,等.基于纳米银颗粒团聚反应的表面增强拉曼光谱法测定牛奶中三聚氰胺的含量[J].食品与发酵工业,2019,45(15)：256 - 261.

[245] Hruzikova J，Milde D，Krajancova P，et al. Discrimination of cheese products for authenticity control by infrared spectroscopy[J]. Journal of Agricultural and Food Chemistry，2012，60(7)：1845 - 1849.

[246] Zhang Z Y. Rapid discrimination of cheese products based on probabilistic neural network and Raman spectroscopy[J]. Journal of Spectroscopy，2020，2020：8896535.

[247] Fourie E，Aleixandre-Tudo J L，Mihnea M，et al. Partial least squares calibrations and batch statistical process control to monitor phenolic extraction in red wine fermentations under different maceration conditions[J]. Food Control，2020，115：107303.

[248] Srivastava S, Mishra G, Mishra H N. Probabilistic artificial neural network and E-nose based classification of *Rhyzopertha dominica* infestation in stored rice grains[J]. Chemometrics and Intelligent Laboratory Systems, 2019, 186: 12-22.

[249] 张阳阳, 贾云献, 吴巍屹, 等. 概率神经网络在车辆齿轮箱典型故障诊断中的应用[J]. 汽车工程, 2020, 42(7): 972-977.

[250] Tuccitto N, Bombace A, Torrisi A, et al. Probabilistic neural network-based classifier of ToF-SIMS single-pixel spectra[J]. Chemometrics and Intelligent Laboratory Systems, 2019, 191: 138-142.

[251] Tsuji T, Nobukawa T, Mito A, et al. Recurrent probabilistic neural network-based short-term prediction for acute hypotension and ventricular fibrillation[J]. Scientific Reports, 2020, 10: 11970.

[252] Almeida M R, de Souza L P, Cesar R S, et al. Investigation of sport supplements quality by Raman spectroscopy and principal component analysis[J]. Vibrational Spectroscopy, 2016, 87: 1-7.

[253] 杨嘉, 丁娟芳, 周元元, 等. 高效液相色谱法快速测定乳粉中的三聚氰胺与双氰胺[J]. 食品科学, 2014, 35(6): 172-175.

[254] Yang Q Q, Liang F H, Wang D, et al. Simultaneous determination of thiocyanate ion and melamine in milk and milk powder using surface-enhanced Raman spectroscopy[J]. Analytical Methods, 2014, 6(20): 8388-8395.

[255] Santos P M, Pereira-Filho E R, Rodriguez-Saona L E. Rapid detection and quantification of milk adulteration using infrared microspectroscopy and chemometrics analysis[J]. Food Chemistry, 2013, 138(1): 19-24.

[256] 王海燕, 桂冬冬, 沙敏, 等. 拉曼光谱结合模式识别算法用以牛奶制品智能判别与参数优化[J]. 中国奶牛, 2018(2): 55-60.

[257] 李华, 王菊香, 邢志娜, 等. 改进的K/S算法对近红外光谱模型传递影响的研究[J]. 光谱学与光谱分析, 2011, 31(2): 362-365.

[258] Wang S J, Liu K S, Yu X J, et al. Application of hybrid image features for fast and non-invasive classification of raisin[J]. Journal of Food Engineering, 2012, 109(3): 531-537.

[259] Zou H Y, Xu K L, Feng Y Y, et al. Application of first order

derivative UV spectrophotometry coupled with H-point standard addition to the simultaneous determination of melamine and dicyandiamide in milk[J]. Food Analytical Methods, 2015, 8(3): 740 - 748.

[260] Fernández Pierna J A, Vincke D, Baeten V, et al. Use of a multivariate moving window PCA for the untargeted detection of contaminants in agro-food products, as exemplified by the detection of melamine levels in milk using vibrational spectroscopy[J]. Chemometrics and Intelligent Laboratory Systems, 2016, 152: 157 - 162.

[261] Capuano E, Boerrigter-Eenling R, Koot A, et al. Targeted and untargeted detection of skim milk powder adulteration by near-infrared spectroscopy[J]. Food Analytical Methods, 2015, 8(8): 2125 - 2134.

[262] Chapelle O, Vapnik V, Bousquet O, et al. Choosing multiple parameters for support vector machines[J]. Machine Learning, 2002, 46(1): 131 - 159.

[263] Saini L M, Aggarwal S K, Kumar A. Parameter optimisation using genetic algorithm for support vector machine-based price-forecasting model in National electricity market[J]. IET Generation, Transmission & Distribution, 2010, 4(1): 36 - 49.

[264] Domingo E, Tirelli A A, Nunes C A, et al. Melamine detection in milk using vibrational spectroscopy and chemometrics analysis: A review[J]. Food Research International, 2014, 60: 131 - 139.

[265] Wong T T. Parametric methods for comparing the performance of two classification algorithms evaluated by k-fold cross validation on multiple data sets[J]. Pattern Recognition, 2017, 65: 97 - 107.

[266] Yang L, Wei F, Liu J M, et al. Functional hybrid micro/nanoentities promote agro-food safety inspection[J]. Journal of Agricultural and Food Chemistry, 2021, 69(42): 12402 - 12417.

[267] Wang X, Zhao J J, Zhang Q, et al. A chemometric strategy for accurately identifying illegal additive compounds in health foods by using ultra-high-performance liquid chromatography coupled to high resolution mass spectrometry[J]. Analytical Methods, 2021, 13(14):

1731-1739.

[268] Hussain A, Pu H B, Sun D W. SERS detection of sodium thiocyanate and benzoic acid preservatives in liquid milk using cysteamine functionalized core-shelled nanoparticles[J]. Spectrochimica Acta Part A, Molecular and Biomolecular Spectroscopy, 2020, 229: 117994.

[269] Yang Z J, Zhang R, Chen H, et al. Rapid quantification of thiocyanate in milk samples using a universal paper-based SERS sensor[J]. The Analyst, 2022, 147(22): 5038-5043.

[270] 吴棉棉,李丹,陆峰. 功能化 SERS 纸基应用于牛奶非法添加物的分离与检测[J]. 光谱学与光谱分析,2016,36(S1):241-242.

[271] Lubes G, Goodarzi M. Analysis of volatile compounds by advanced analytical techniques and multivariate chemometrics[J]. Chemical Reviews, 2017, 117(9): 6399-6422.

[272] Takamura A, Ozawa T. Recent advances of vibrational spectroscopy and chemometrics for forensic biological analysis[J]. The Analyst, 2021, 146(24): 7431-7449.

[273] Zhang Z Y, Jiang M Q, Xiong H M. Optimized identification of cheese products based on Raman spectroscopy and an extreme learning machine[J]. New Journal of Chemistry, 2023, 47(14): 6889-6894.

[274] 李靖,李梦银,陈守慧,等. 牛乳主要过敏原的拉曼光谱检测分析[J]. 食品工业,2021,42(12):242-246.

[275] Xiouras C, Cameli F, Quilló G L, et al. Applications of artificial intelligence and machine learning algorithms to crystallization[J]. Chemical Reviews, 2022, 122(15): 13006-13042.

[276] Wu S J, Cui T C, Li Z, et al. Real-time monitoring of the column chromatographic process of *Phellodendri Chinensis Cortex* part I: End-point determination based on near-infrared spectroscopy combined with machine learning[J]. New Journal of Chemistry, 2022, 46(19): 9085-9097.

[277] Xiao D, Li H Z, Sun X Y. Coal classification method based on improved local receptive field-based extreme learning machine algorithm and visible-infrared spectroscopy[J]. ACS Omega, 2020, 5(40): 25772-

25783.

[278] Xu Y, Zhong P, Jiang A M, et al. Raman spectroscopy coupled with chemometrics for food authentication: A review[J]. TrAC Trends in Analytical Chemistry, 2020, 131: 116017.

[279] Mozhaeva V, Kudryavtsev D, Prokhorov K, et al. Toxins' classification through Raman spectroscopy with principal component analysis[J]. Spectrochimica Acta Part A, Molecular and Biomolecular Spectroscopy, 2022, 278: 121276.